Serbia on the Road to EU Accession

Consequences for Agricultural Policy and the Agri-food Chain

Serbia on the Road to EU Accession

Consequences for Agricultural Policy and the Agri-food Chain

Siemen van Berkum

LEI, part of Wageningen UR

Natalija Bogdanov

University of Belgrade, Faculty of Agriculture

www.cabi.org

CABI is a trading name of CAB International

CABI
Nosworthy Way
Wallingford
Oxfordshire OX10 8DE
UK

Tel: +44 (0)1491 832111
Fax: +44 (0)1491 833508
E-mail: cabi@cabi.org
Website: www.cabi.org

CABI
875 Massachusetts Avenue
7th Floor
Cambridge, MA 02139
USA

Tel: +1 617 395 4056
Fax: +1 617 354 6875
E-mail: cabi-nao@cabi.org

A catalogue record for this book is available from the British Library, London, UK.

Library of Congress Cataloging-in-Publication Data

Berkum, S. van
Serbia on the road to EU accession : consequences for agricultural policy and the agri-food chain / Siemen van Berkum, Natalija Bogdanov.
p. cm.
Includes bibliographical references and index.
ISBN 978-1-78064-145-4 (alk. paper)
1. European Union--Serbia--Membership. 2. Agriculture and state--Serbia. 3. Serbia--Economic conditions--21st century. 4. Food supply--Government policy--Serbia. I. Bogdanov, Natalija Lj. II. Title.

HC240.25.S4B47 2012
338.1'84971--dc23

2012005598

ISBN-13: 978 1 78064 145 4

Commissioning editor: Claire Parfitt

Printed and bound in the UK from copy supplied by the authors by CPI Group (UK) Ltd, Croydon, CR0 4YY.

Contents

Acknowledgement

This book is based on an initiative of both the Serbian and Dutch government to commission a research team with members from both Serbia and the Netherlands to execute a study on the implications of the EU accession for the Serbian agri-food supply chain. We thank those who were involved in formulating the terms of reference and in providing the financial support for the study: Jan Deelstra (director International affairs), Adri de Roo (agricultural counsellor in Zagreb/Belgrade) and Tim Heddema of the Dutch Ministry of Economic Affairs, Agriculture and Innovation (EL&I) and Milos Milovanovic (assistant minister) of the Serbian Ministry of Agriculture, Forestry and Water Management (MAFWM). Both Ministries established a guiding committee with members of their staff, to which we like to express our gratitude for their comments on the report and the useful guidance during meetings in both The Hague and Belgrade.

In particular we want to thank Ate Oostra and Gerrit Meester, both former employees of the Dutch Ministry of EL&I for their inspiring support to establish the international cooperation between the two countries and their sustained interest and professional support during the execution of this project.

The outline and approach of this book are greatly inspired by 'Turkey in the European Union. Implications for Agriculture, Food and Structural Policy', edited by Burrell and Oskam (2005). One of the authors of this book contributed to that Turkey study. We acknowledge the benefits we have had from the Burrell and Oskam study. Although this study very much follows the approach of the report on Turkey, it has its own accents and elaborations on topics of interest in the Serbian case.

Various people contributed to the final outcome of the report by providing information and advice and by commenting on draft chapters. We acknowledge the constructive remarks by Tim Heddema, Arie Oskam, Ate Oostra, Peter Nowicki, Ida Terluin, Gloria Solano (also for major contributions to Chapters 2 and 9) and Jo Wijnands. We are also thankful to Ivana Ivkov, Bojan Zivadinovic, Marina Rajkovic (employees of Serbian MAFWM) and Milan Prostran (Serbian Chamber of Commerce) who provided data and information, and useful suggestions to improve the analusis in talks and interviews. Of course the authors remain responsible for the content, as well as any errors and omissions.

Siemen van Berkum and Natalija Bogdanov
The Hague, Belgrade
January 2012

Summary

Serbia officially applied for European Union (EU) membership on 22 December 2009. This report aims at providing relevant background information, in the event that a decision is made to begin accession negotiations with Serbia. If this happens, then the full set of Copenhagen criteria will become relevant. This report focuses on one of the Copenhagen criteria: the existence of a functioning market economy, including the existence of institutions required to support such a market economy within the EU. The report focuses on this aspect of the Copenhagen criteria in relation to agriculture, food, rural development and structural policy. A key question is what needs to be done to prepare the Serbian agri-food sector for EU membership. The budgetary implications of accession are also addressed.

Serbia's economic situation

Serbia is a country located in the central south-eastern part of Europe with a population of 7.3 million people. The average per capita income was 4,100 euro in 2009, which is approximately 17 per cent of the EU-27 level. Corrected for purchasing power differences, per capita GDP is about one third of the EU-27 average.

The sectoral distribution of employment is relatively stable over time: services account for the largest share (49 per cent), and industry and the agricultural sector for 25 per cent and 26 per cent, respectively, in 2008. Unemployment rates are high even compared to other transition economies of the region.

The trade deficit has been growing in most years during the last decade, accounting for 19 per cent of GDP in 2009. Serbian trade is in deficit with all regions except for the countries of CEFTA. Serbia's current account deficit had been comfortably balanced by significant foreign investment. However, since 2007 this source has been declining and a current account deficit has to be compensated with debt-creating capital inflows. Total external debt in Serbia reached 75 per cent of GDP at the end of 2009, reflecting a considerably higher level of indebtedness than the average of upper middle income countries.

The richest 20 per cent of households in Serbia received around 38 per cent of total disposable income in 2008. The inequality in urban areas is higher than in rural areas. Poverty in Serbia decreased during the last decade: in 2002 about 14 per cent or 1 million people were classified as poor, against 7 per cent or 0.5 million people in 2007. The regions of Belgrade and Vojvodina account for 66 per cent of total GDP. In per capita terms GDP in the Belgrade region is twice as much as in Vojvodina, 2.5 times higher than in Sumadija and Western Serbia, and three times the level in Southern and Eastern Serbia. The regional income inequalities arise mainly because of the differences in type of economic activities. Poorer regions generally have a bigger share of resources in agriculture.

Institutional background

The institutional system in Serbia during the 20th century has been determined by frequent changes of state borders, along with the socio-economic and political system. These changes have brought radical shifts in the legal, organizational and implementation mechanisms. As in other communist countries, after World War II, the institutional system was deeply committed to the ruling political ideology. An important feature of the institutional structure was the fact that Vojvodina developed its own institutions at the time of the Austro-Hungarian monarchy. As a consequence, institutions are better organized and have a longer tradition in this part of the country.

After the political changes in 2000 substantial reforms of institutions and legislation began. In the field of agriculture a major challenge at the beginning of transition was combating the grey economy and establishment of an efficient system of border controls. Fundamental reform of laws and institutions related to land, cooperatives, advisory services, laboratory systems and so on took place slowly, due to a lack of inter-ministerial cooperation and extensive overlapping of the responsibilities between different institutions. As a result substantial structural reforms in the sector were delayed.

In view of EU membership, the Serbian government has been working on the adoption of administrative arrangements adjusted to EU standards, whereas administrative and social capacities developed slowly. The knowledge and innovation system (extension, education, research and development) has been reformed, but is still not efficient enough and is inadequate for the needs of small farms. Producer associations and associations, horizontal and vertical networking of the business sector are insufficiently and unevenly developed.

Significant progress has been made on the side of institutional building and preparation of strategic documents related to the use of pre-accession EU funds. On the other hand, the major constraint remains the lack of competent staff. Enhancing skills of human resources of the administration is certainly a priority, as well as better coordination between the various governmental bodies.

Primary agricultural production

Primary agriculture accounts for 10 per cent of Serbia's GDP, about a quarter of the country's employment and 23 per cent of merchandise exports. About half a million people work in Serbian agriculture. The total agricultural area is about 5 million hectares, of which 65 per cent (3.3 million hectares) is arable land.

Farm and land structures are fragmented, yet with considerable regional variety. Farms in Vojvodina are larger than in Central Serbia and the Southern regions of the country, where (semi-)subsistence farming prevails.

Farm land has been privatized, yet there is still much controversy about the ownership of previously nationalized or confiscated land. The state still has a significant land area that can be rented by individual farmers. However, lack of clear ownership rights for a significant part of the land hinders an appropriate land market operation and, hence, reduces an efficient use of land.

Agricultural production is dominated by crop production. Maize is the major crop of the country, followed by wheat, oilseeds (sunflower and soybeans) and vegetables. Yields of cereals, sugar beet, rape and sunflower seeds as well as of major vegetables and fruits are still below levels of the pre-1990s. The most important livestock products are pork, milk and beef, for which current yields are also lower than in the 1980s. Compared to EU countries, yields in Serbia are in the lower end of the spectrum.

Agricultural prices in Serbia are lower than in neighbouring countries, indicating the country is price competitive. Data on labour productivity are scarce, yet indicate that the average agricultural income is lower than in non-agricultural sectors.

Rural population and agricultural workforce

Serbia's total population is decreasing over the last decade. Population in most urban municipalities has increased, while the population number declined in most of the rural municipalities.

As a result of rather poor education and training, the knowledge and skills of the rural labour force are generally inappropriate for the requirements of modern technology; this adversely affects the overall capacity of the labour force in rural areas. This low quality of the workforce can be considered as one of the factors hampering economic development in rural areas.

The unemployment rate in the rural labour market in Serbia is high, especially of the younger, higher educated population and women. The economic crisis has highlighted the unfavourable trends in the rural labour market.

Nearly half of the rural labour force is employed in the agricultural sector and about one third in the services sector. Agriculture remains the predominant activity in most rural areas, characterized by small farm units, low productivity and low farm incomes. A large number of farms are subsistence farms with a very small surplus production for the market. The findings of the research on social inclusion of the rural population have shown that in 2010 most of the farms were included in the category of subsistence farming (65 per cent), while only 35 per cent of the farms are involved in farming production for the market.

Poverty in Serbia is more widespread in rural areas than in urban areas and is strongly linked to the performance of the agricultural labour force and the rural labour market, as one half of all those employed in agriculture lives in poor households. Poverty is more prevalent in southern and hilly/mountainous regions in comparison to northern plain regions.

The agri-food chain

The structural features of the agri-food supply chain in Serbia point at a dual structure in every level of the chain, in which typically a large number of small to very small farms and firms dominate. Next to that there are farms and companies with medium to large scale operations. At a primary level, farm structures are highly fragmented with a large part of the sector being (semi-)subsistence. The structure of the food processing industry varies across branches, but overall there are only a few companies which account for a major share of turnover. Market power, however, does not seem to exist, although hard evidence to verify this is not available.

Home consumption and direct sales of food (the informal economy) are highly significant in Serbia. This implies that a large part of the agricultural sector is not integrated in the agri-food supply chain, either by contracting or by (any other means of) sustainable commercial relationships. These are the large group of (very) small-scale farmers. Next to that, there is a large number of small processing units. Their position is critical as they have to comply with the requirements of an increasingly demanding retail chain. These requirements are driven by the coinciding forces of increasing consumer awareness and public policies aiming at establishing the legal framework for food safety and quality standards. Even the larger enterprises have difficulties in adapting to the European requirements, given the small number of licences for export to the EU in the meat and dairy sector. Foreign investment in the retail as well as in the food processing industry may have an important role in the restructuring and modernization of Serbia's food supply chain, but investors only come in under the condition of sufficient political and economic stability.

The food chain in Serbia is mainly governed by free market principles with arm's length control of price formation. This holds especially for the relationship between the primary and the processing levels of the supply chain, although several branches of the processing industry work with contractual agreements for the supply of raw materials. The modern retail chain in Serbia increasingly demands guarantees on quality and food safety at competitive prices. These guarantees are enforced by contracts. A major challenge for the farming and processing sector is, then, to meet the growing requirements and standards all along the chain. Failure to meet such requirements may result in increasing import penetration.

Agricultural, food, rural and structural policies

Agricultural policy reform began after the political changes of 2000. An important policy objective was to increase the volume of production, which was at a very low level after decades of isolation. The main instruments of support were based on production (premiums and direct payments related to yields) as well as reimbursements for the inputs used. Thanks to these incentives production increased significantly within a short period and exports increased. The main instruments of market policy were high customs duties for products for which the country was thought to become competitive and export subsidies. Market intervention purchases and price support policy were not used.

In 2005, the government adopted the Agriculture Development Strategy, which was replaced in 2010 by the Agriculture Development Strategy for 2010-2020. This strategy is implemented through the National Agricultural Programme of the Republic of Serbia (for 2010-2013) and the National Rural Development Programme of the Republic of Serbia (for 2011-2013).

Agricultural budgets fluctuated significantly over the past years, both in total outlays and for each major policy instrument. This is a sign that agricultural policy was in transition and that no long-term framework existed. The spending of financial resources was not planned or specified in advance, and changed depending on the annual state budget and the share of agriculture in it. The lack of consistent policy, i.e. clearly defined preferences and instruments for their implementation, caused both inefficiencies and overall policy impacts that were below the expected and objectively possible level.

Although changes in the level of support were painful for agricultural producers, the frequently radical changes in the type and form of payment were perhaps even more problematic. In 2004 a system of registration of holdings was introduced that gradually became a prerequisite for the use of public support. Terms of eligibility for support became more complex, and were to the advantage of the large producers. The unpredictability of the support system gives producers a sense of uncertainty, which discourages them from investing in their farming. Half of all farms were registered as potential beneficiaries of budget support. Yet, presently only 10-15 per cent of households use direct support per hectare, and only for arable land.

Market support measures and direct support for producers prevail as policy instruments. A receipt of input purchase is sufficient for receiving support, instead of meeting standards of quality, environmental protection and/or animal welfare requirements as in the EU. Market support measures and direct support for producers were particularly inconsistent and without a clear concept in the livestock production.

Support for rural development has become more actively implemented since 2004. Most of the funds used support improving farm competitiveness. The basic mechanism is an interest rate subsidy on loans for farm mechanization and equipment. Some elements of support are based on the model of the EU's Less Favoured Areas (LFA) directive, but without clearly defined rules. In recent years there have been some attempts to deal with sensitive social issues through support for rural development (support for the Roma population in rural areas, for refugees, for non-commercial farms, etc.), but this practice was later abandoned.

The Serbian agricultural policy requires fundamental reforms at all levels. The model of support for agriculture that is applied in Serbia is significantly different from EU practices, and will have to adjust if Serbia should become a member of the EU. Adjustment will require new (accredited) institutions, greater transparency and access to support for a wider range of agricultural producers.

Serbia's foreign trade position

Serbia is a net importing and open economy with foreign trade as a significant part of GDP. The EU is by far the country's major trading partner, both for all merchandise trade and for trade in agricultural products. Serbia is a net exporter of agricultural products, especially of fruit and vegetables, cereals and sugar. Next to the EU, Serbia's main export markets are ex-Yugoslavian republics (especially Bosnia/Herzegovina and Montenegro). Trade figures show that Serbia has a strong competitive position in the West Balkan region.

Serbia is close to the finalization of its negotiation for WTO membership. That membership would integrate the country further into the global economy. The compliance with WTO rules is well under way, although there are still some critical issues to be tackled with respect to import licensing, sanitary and phytosanitary issues, intellectual property rights and services. A consequence of becoming a member of the WTO is that Serbia's import tariffs on agricultural products will decline to offer more market access to trade partners who will have to do the same. On the other hand, although current MFN import tariffs are still significant, they have no or only very little impact on trade with major trading partners due to Serbia's many free trade agreements, most notably under the CEFTA and the Interim Trade Agreement as part of the Stabilization and Association Agreement (SAA) with the EU.

Trade between Serbia and the EU is free of tariffs and quantitative restrictions for many agricultural products. Nevertheless, there are also a substantial number of products for which Serbia will apply tariff-rate quotas (TRQs) and/or tariffs in the years to come while there are also categories of products for which import duties are subject to gradual reduction over a period of 6 years. This implies that customs protection will remain at least for some years for a substantial group of products. Serbian exports to the EU are free of any duty or levy, except for baby beef, sugar and wine, which are subject to TRQs. Except for the categories mentioned, Serbian products have free access to the EU, but whether they will benefit from the opportunities offered depends on their competitiveness in terms of price and quality. In addition, exports may face difficulties due to a lack of legal harmonization and delay in the introduction of necessary standards in the field of food safety, traceability and quality along the whole supply chain. To date, Serbian standards and regulations are not yet fully harmonized with those of the EU.

Environment and agriculture

The main agricultural sources of water pollution are the uncontrolled effluents of livestock farms (e.g. animal manures), crops, slaughterhouse wastes, chemical fertilizers and pesticides. Fertilizer and pesticide use in Serbia is relatively small compared to most EU countries. Soil fertility is reduced by wind and water erosion, and poor drainage. The lack of systematic soil monitoring and analysis hampers appropriate insights into the country's soil status, and consequently a strategy about what to do to improve soil fertility.

Organic farming has great potential in Serbia, which is to some extent already used for the export market. Farmers use low levels of chemical inputs, which implies that conversion periods for organic certification are short. However, the potential for growth of this sector is limited by a combination of little government support, capital constraints and poor supply chain organization, which leads to inefficient production, processing and marketing. Suggestions to improve the organic farming sector focus on credit facilities, education, promoting the use of modern technology and propagation material, and support for associations and cooperatives.

Serbia's territory is very rich in biodiversity. The main threats coming from agriculture are natural habitat destruction by using relatively pristine areas for agricultural purposes and especially through applying monocultures. Over-exploitation and overuse of agro-chemistry and agro-machinery are uncommon in Serbia, and hence do not affect biodiversity very much for the moment.

The analysis of environmental impacts of agriculture in Serbia encounters two important limitations. First, there is a lack of data, or data available for only over a short period. Second, the policy and institutional framework has been established only very recently or is still in development, which makes a satisfactory policy impact assessment unfeasible.

Serbia has demonstrated its interest in and commitment to protecting the environment and the country's biodiversity with the ratification of international agreements and the adoption of a national legal framework. However, some EU directives in the field of the environment (e.g. the Nitrates Directive) have not been implemented yet. In the support system for agricultural production in Serbia, cross-compliance measures with environmental, animal welfare, public, animal or plant health standards do not yet exist.

Serbia is facing difficulties in implementing policies and strategies for adequate environmental protection. Besides a lack of financial resources that are necessary to provide farmers with incentives to reach the policy goals, the national legal framework is of (very) recent date and lacks appropriate institutional structures for implementation and enforcement of regulations. Central collection of data and monitoring of environmental indicators are missing or fragmented among diverse projects. Finally, investments in agriculture in irrigation systems, drainage and flood protection, or in waste management systems (in slaughterhouses) would contribute much to reaching environmental protection targets. Yet, the current economic instruments in policies to encourage farmers and other actors in the agri-food supply chain may be too weak to get actual investment started.

Animal and plant health

Much legislation and many procedures have recently been transposed into rules and guidelines for all participants in the agri-food chain. Yet, legislative and administrative changes alone do not guarantee a proper implementation of the veterinary and phytosanitary *acquis* of the EU. The implementation of regulations with the rigour and expertise necessary for them to function needs appropriate institutions and effective organizations, and the time to build up the expertise in this area.

The small-scale and fragmented structure of Serbia's agriculture, both at the primary, the processing and the trade level, makes monitoring and control of animal and plant diseases difficult. Furthermore, investments in production facilities and methods that help to achieve higher standards fall short of what is necessary in order to comply with EU requirements.

There are, however, important differences between the plant and animal sectors, with the phytosanitary standards reached in the plant sector being conformable with those required by international trade. In contrast, the animal sector faces more difficulties. Given the highly infectious nature of List A and some List B diseases (a number of which remain endemic in Serbia), the potential trade consequences of an outbreak of one of these diseases are catastrophic, involving a ban on all imports from the source country or region until the disease has been eradicated according to the protocols agreed within the OIE, and international confidence has been restored.

Currently, Serbia's animal sector is only very little involved in international trade and it is envisaged that animal products would not be competitive on the EU market. Therefore, the likelihood of Serbia becoming a significant exporter of animal products is small. Still, even with small export flows to the EU, the country's health status should be indisputable and up to the standard of EU requirements in order to maintain the confidence of trading partners in EU's animal products. Serbia's animal sector does not have that status yet.

Budgetary consequences of Serbia joining the EU

Taking into account how the presently on-going discussions on CAP and EU budget framework reforms are developing, and including some assumptions with regard to economic growth and exchange rate developments, the budgetary consequences of Serbia joining the EU within five to seven years would be 1.1 to 1.67 billion euro. The range is due to the different time frame and the options of phasing in or not phasing in direct payments. If Serbia enters within seven years, the net budget revenue for Serbia will be 45 million euro less compared to joining within five years, as the country's contribution to the EU budget will be higher in the first case. The amounts are about 2.6 to 3.7 per cent of Serbia's GDP over five or seven years respectively, which is a significant injection into the economy.

Consequences for the EU of Serbia's accession

While the impact of EU membership on the Serbian economy is expected to be substantial, consequences for the countries that are already EU members at the time that Serbia joins are assumed to be relatively small. This is based on the relative small size of the Serbian economy compared to the EU-27 (which is only 0.25 per cent of the EU-27 total GDP) and the rather limited transfer of money from EU members to Serbia (less than 0.1 per cent of

EU-27 GDP). Furthermore, Serbia's agricultural production volumes are minor compared to EU-27 levels, and there is little redirection of trade flows due to Serbia becoming an EU member. Therefore, given these aspects mentioned, the entrance of Serbia will provide no reason for policy reforms as such.

Implications for agricultural and rural policies and for the agri-food sector

EU membership will have major implications for Serbia's agricultural policy and the sector's structure. At the present time, Serbia's agricultural policy differs significantly from EU's CAP, as the instruments used, the criteria for and level of support are different. Subsequently, aligning with EU policy will have significant distributional effects, not only among farmers when the criteria for and levels of support change, but also between the agricultural and non-agricultural sector. Furthermore, the full implementation of the *acquis* on agriculture, food safety, veterinary and phytosanitary issues, and environmental measures to avoid adverse effects from agricultural activities all require investments at farm and firm level. IPARD (EU's Instrument for Pre-accession Assistance in Rural Development) and subsequently rural development funds (when a member of the EU) are available to adapt to the requirements set by the *acquis*. Nevertheless, part of the primary and processing sector will find itself too small economically to make these investments. Consequently, the sector will face a restructuring and consolidation process. While this process increases efficiency and, hence, competitiveness of the sector, and thus is beneficial from the sector's overall perspective, there are possibly great social consequences, because finding alternative employment for those leaving the agricultural sector may be problematic. Economic growth that creates employment opportunities in other sectors is an important condition for a smooth process of structural change in the agricultural sector.

Challenges and how to take them up

Serbia faces many challenges should the country become an EU member in the medium term. A major condition for EU membership is the adoption and implementation of the set of rules, laws and regulations that is the *acquis communautaire* of the European Union. To date, the fundamental reform of laws and institutions has been too slow, especially with respect to land, cooperatives, advisory services, the food safety system and so forth. Substantial structural reforms in the sector were delayed as a result. Speeding up the process of adopting legislation and establishing the necessary institutions to implement the legislation is required to provide the agri-food chain a well-founded legal and institution context in which to do business.

In view of EU membership, the Serbian government has been adopting many administrative arrangements, whereas administrative and social capacities to implement these arrangements are still deficient. Enhancing human resources of the administration should be a priority, as well as a better coordination of tasks and responsibilities between the various governmental bodies. Several twinning projects have been initiated to support capacity improvements, such as in the area of food safety and farm data (linked to FADN). Intensifying the process of capacity building is recommended, because improving the competences of the public administration is a crucial element in the process of preparing for EU membership in the years to come.

The CAP will remain the principal framework for the agricultural policy in the EU in the medium term. Aligning agricultural policies with the CAP will, therefore, facilitate eventual accession to the EU. Serbia needs to be well informed about the way the CAP evolves during the process of bringing its policies into line with those of the EU. Such information is important in order to evaluate the possible impact of policy changes on the agricultural sector's development, and thus to design appropriate policy answers to prepare the sector for the effects identified.

Next to changes in direct payments' conditions the CAP moves to rural development (second pillar) measures. That by itself is a reason to give ample attention to that policy area in Serbia's process of preparing for EU accession. Yet, effective rural development programmes are also needed to ensure that the agricultural sector's modernization is balanced and equitable. Serbia's agricultural sector has a dual structure with many small-scale, subsistence farms, and regional differences in economic development are considerable. Restructuring and modernization of the agri-food sector, hence, may have significant regional economic effects. It is important therefore to explore the non-agricultural potential of rural areas and to design territorial, rural development strategies. Local authorities and stakeholders should be involved in designing these programmes.

A major task for the Serbian government is to provide sufficient public services that help the agri-food sector to prepare for EU membership. As said beforehand, the Serbian agri-food supply chain may expect increasing competition if Serbia joins the EU internal market. Advisory (extension) services, agricultural information services, and agricultural education and research are public services with a proven valuable impact in helping to adapt the Serbian agri-sector enterprises to changing policy and market conditions. Currently these services fall short of what the sector needs. Increasing public investments in these areas would enhance the sector's capacity to stand the increasing competitive pressures it will encounter when Serbia is a member of the EU.

Chapter 1

1 Introduction

Serbia officially applied for European Union (EU) membership on 22 December 2009. The government of Serbia has set a goal for EU accession in 2014 (EurActiv, 2010). The European Commission has backed this initiative, but has indicated that Serbia's EU accession could take five to seven years to accomplish.[1] When becoming an EU member the applicant needs to take on board the entire body of EU legislation.

This report aims at providing relevant background information in the event that a decision is made by the EU to begin accession negotiations with Serbia. If this happens, then the full set of Copenhagen criteria will become relevant. This report focuses on one of the Copenhagen criteria: the existence of a functioning market economy, including the existence of institutions required to support such a market economy within the EU. The report focuses on this aspect of the Copenhagen criteria in relation to agriculture, food, rural development and structural policy.

This chapter addresses the background, approach and structure of the study.

1.1 Background

Advancing European integration is a stated priority of the Serbian Government, and major progress has been made in this area with the signing of a Stabilization and Association Agreement (SAA) on 29 April 2008. European integration is an objective supported by almost two thirds of Serbian society. Nevertheless, the political situation and the EU accession process in Serbia remain influenced by recent history. While the SAA has been signed and ratified in the Serbian Parliament, the EU delayed its entry into effect and required that Serbia should demonstrate closer cooperation with the International Criminal Tribunal for the former Yugoslavia (ICTY). Only in early December 2009 did the EU unfreeze the interim trade agreement -- the trade part of the SAA, which was already unilaterally implemented by Serbia in January 2009. In June 2010, the EU Council decided to start the ratification process of the SAA.

On 22 December 2009 Serbia submitted its application for EU membership to the Swedish Prime Minister, who was acting in his role of the EU Presidency. The European Council made Serbia a candidate for membership on 1 March 2012 (General Secretariat of

[1] According to European Commission Vice-President Barrot in UPI.com World News, 'European Union wants Serbia to join', 26 December 2009. This may even be an optimistic estimation. Croatia's accession process may be taken as an example: the country's application for membership was submitted in 2003 and accession negotiation talks finished on 30 June 2011. The ratification process by parliaments of all 27 EU member states is expected to be concluded by June 2013. This implies that Croatia may join the EU by July 2013 (statement of President Barrosso, 10 June 2011, MEMO/11/397).

the Council, 2012). The next step in the procedure is that the EU Council has to decide on when accession talks may start. Next to political conditions Serbia has to satisfy economic requirements in order to meet the obligations of membership. With respect to the latter, the Copenhagen criteria refer to '*the existence of a functioning market economy as well as the capacity to cope with competitive pressure and market force within the Union*'. (Bulletin, EC 6 - 1993: 13).

Agriculture and its related industries play an important role in the Serbian economy, as the sector accounts for around 20 per cent of GDP. This report focuses on the ability of the sector to comply with the requirements set by the Copenhagen criteria and the *acquis communautaire* in relation to agriculture, food, rural development and structural policy. As the present EU member states are discussing the future of the Common Agricultural Policy (CAP), for which the European Commission presented its outline in November 2010,[1] the *acquis* related to the agricultural and food sector may change in due course. It is against this background that the Serbian agricultural sector and the relevant institutions have to adapt and transform in order to join EU's internal market.

1.2 Objective and questions

The main objective of this study is to provide a balanced overview of Serbian agriculture, food and rural areas, taking Serbian accession to the European Union in the medium term (say 5 to 7 years) as a working hypothesis. Consequences for both Serbia and the EU are explored mainly on the basis of available literature, databases and information provided by stakeholders.

The report attempts to answer the following questions:

- What is the economic situation of Serbia and its regional income structure?
- What are the main economic institutions involved in agriculture and food in Serbia; and how is agricultural, rural and structural policy organized?
- What is the current performance of the Serbian primary agricultural sector?
- What are the structural features of the primary sector?
- What are the structural features of the agri-food chain and how can linkages between chain actors be characterized?
- What are the main differences between Serbian and EU agricultural policies?
- What is the trade position of Serbia, with respect to the EU and in a regional perspective?
- What is the relationship between agriculture and the environment in Serbia; and how does this compare with EU standards?
- What standards apply in Serbia with respect to food safety, plant health, animal health and biosecurity?
- What are the implications, for both Serbia and EU member states, of applying EU agricultural, food, rural and structural policy rules to Serbia? Here budget implications play an important role.

[1] Further discussion on the EC Communication on the CAP's future in chapter 11.

- What are the main challenges Serbia faces for EU membership and how should it take them on?

In this research, we focus on Serbia[1] and the EU-27, sometimes divided into EU-15, NMS (the ten new member states from 1 May 2004) and AC-2 (Bulgaria and Romania). Although Croatia is likely to be an EU member already if Serbia enters, we do not include this country in our analysis at this stage for, within the deadline set, a complete set of relevant information is unavailable.

1.3 Approach and structure of the study

This report will provide a description of the agricultural supply chain and its institutional environment from an economic perspective. Such a description will be used as a base for the assessment of the possible impact of EU membership on Serbian agriculture. The description and further analyses will be largely based on available literature and data. There is not much up to date literature on Serbian agriculture available in English, German or French, neither from sources such as international organizations like the EU, World Bank, OECD and FAO or from Serbian sources such as the Ministry of Agriculture, universities, research institutes and Serbian authors. Next, earlier experiences of the research team of this study indicate that socio-economic data on the agri-food supply chain can be a problem as well.[2] Some data on primary level/agriculture can be found at websites and in reports of the Statistical Office of the Republic of Serbia, but where data on production levels, crop areas, livestock numbers and trade might be readily available, economic data on upstream and downstream agri-food industries are rather limited. Due to this scarcity of data and information sources, the study also entails interviews with several experts.[3]

In addition to a lack of statistical data on the agri-food supply chain, there is also very little analysis done so far on the consequences of Serbian accession to the EU . The studies considering this topic take into account snapshots, yet do not provide an overall economic or agricultural sector analysis. The expected economic consequences of Serbia being a member of the EU are very important for determining EU budget allocations, Serbia's absorption capacity for structural funds and opportunities for trade and foreign direct investment. With a lack of references to research done, this study has to be based on several assumptions on general economic growth in Serbia in the medium-term, institutional developments and performances of the agri-food supply chain in the years to come. These assumptions and estimations will be spelled out in more detail in the chapters that focus on the expected consequences of Serbian accession to the EU.

The structure of the report is as follows. Chapter 2 provides general information on the Serbian economy. It provides the macroeconomic background against which we analyse agri-food issues and the consequences of Serbia's accession to the EU. Chapter 3 describes the current institutional framework of Serbian agriculture. Specifically, we examine institutional developments that have implications for the alignment of Serbian agricultural

[1] The territory of Kosovo and Metohija (UNMIK) is not taken into consideration in this report.

[2] The FP7 research project AgriPolicy Network (2008-2010) spent much effort in bringing together key agri-food sector and market information in a systematic and comparable way for the 12 new member states and eight candidate and pre-candidate countries. See www.agripolicy.net.

[3] Interviews with Chamber of Commerce, with University representatives, etc.

institutions with those of the EU, with the aim of providing insights relevant to Serbia's possible accession to the EU and to expectations of future developments. Chapter 4 provides an overview of the agricultural sector in the Serbian economy, covering production, prices and trade.

Chapter 5 concerns the rural population (those living in rural areas) and the agricultural workforce (people working in agriculture) in Serbia. This chapter examines urban-rural differences in labour force participation and unemployment. Education levels are examined, as well as the health and nutritional status, and the incidence of poverty, among the rural population. Chapter 6 gives an overview of the agri-food sector from a chain perspective. In chapter 7, the outlines of the agricultural, food, rural and structural policies are discussed, focusing mainly on the overlap and differences with EU policies. Chapter 8 deals with the trade position of Serbia and its trade relations with its main trading partners. Chapter 9 studies the relationship between agriculture and the environment in Serbia and chapter 10 deals with animal and plant health. The issues covered in chapters 9 and 10 are important not only for agricultural production and trade, but also directly for society as a whole.

Chapter 11 draws on the previous chapters of the report in order to derive the likely consequences for Serbia's agricultural and agri-food sectors should it become an EU member in the future. The key question is: what needs to be done to prepare the agri-food sector for EU membership? Also, budgetary implications of accession are important to address. Here, it is necessary to make a number of assumptions. Our quantitative analysis depends, of course, on the precise numbers that we have assumed in order to describe changes over a period of five years. However, our qualitative analysis and conclusions are often more robust in that they depend on assumptions about general trends and developments that are less controversial. The chapter concludes with the major findings of the study.

Literature

EurActiv.com, *Serbia aims to beat EU entry 'speed record'*. 04 January 2010.

Barrosso, *Statement by President Barroso on Croatia - Commission proposes to close the last 'Chapters' in the accession talks.* MEMO/11/397. Brussels, 10 June 2011.

General Secretariat of the Council, *European Council 1/2 March 2012. Conclusions.* EUCO 4/12, CO EUR2, CONCL1, Brussels, 2 March 2012.

Chapter 2

2 Serbia's economy and regional income distribution

2.1 Introduction

This chapter provides the macroeconomic context within which other chapters analyse agri-food issues and the consequences of the Serbia's accession to the EU. After a brief overview of Serbia and its economy in section 2.2, section 2.3 gives information on the economic growth, inflation, employment, exchanges rates, trade and the balance of payments on the current account and external debt.

Section 2.4 provides information on the size of the economy and section 2.5 analyses inter-household and regional income distribution in some detail, since large regional differences in income in an acceding country have implications for targeting EU structural and rural policy. A related issue is poverty and this section also provides information on the size and importance of poverty in Serbia. Section 2.6 summarizes.

2.2 Serbia at a glance

2.2.1 Country and population

Serbia, officially the Republic of Serbia, is a country located in the central and south-eastern part of Europe and has a population of 7.3 million (Table 2.1). The World Bank classifies Serbia as an upper middle income country with a per capita income of 4,100 euro per annum.

Table 2.1. Serbia's land and population resources.

Land area (1000 ha) a)		**Population b)**	
Total land surface (2008), of which	8,836	Population (2009)	7,320,807 c)
Agricultural land	5,056	Average annual growth 2000-2009 in %	-0.3
Forests	2,618		

Source: a) FAO (2010); b) SORS (2010).
Note: c) Data on Kosovo and Methohija are excluded.

2.2.2 Key economic indicators

Table 2.2. Key indicators of Serbia's economy.

	National accounts(1) (3)		
	2009	% average annual growth 2000-2009	
GDP euro(a) million(1)	29,967	GDP euro million(1)	4.4
GDP per capita (euro)(1)	4,093	GDP per capita(1)	4.7
GDP per capita (PPP[1])(3)(a)	7,175		
	Structure of the economy(1)		
Sectoral shares of GDP, 2009	%	% average annual sectoral growth 2000-2009	
Agriculture	12.1	Agriculture	2.1
Industry	18.3	Industry	-0.5
Services	66.4	Services	6.0
Construction	3.3	Construction	-0.1
	Employment(1) (3) (4)		
Employment by sector, 2008	(%)(3)	Employment indicators, 2009	(%)
Agriculture	25.1	Labour force participation rate(4)	52.2
Industry	22.8	Unemployment rate(1)	16.6
Services	48.7	Labour force growth per year, 2000-2009(4)	-0.9
Construction	3.4		
	Exchange rate and inflation(2)		
	2009		%
Exchange rate RSD/euro (period average)	94.1	Exchange rate index RSD/euro (2000=100) 2009(h)	61.1
Consumer Price Index (2005=100)	144.6	% Average annual IPC change 2005-2009	9.7
	Current account & external debt(2)		
	2009		%
Current account balance (Million euro)	-2,084	% Avg. annual growth of negative balance 2002-09	17.6
% Total external debt/GDP	75%	Current account in GDP %	7.0
		% Average annual external debt growth 2000-09	9.24
	Merchandise trade(1)		
Exports	2009	Imports	2009
Exports (euro million)	5,961.3	Imports (euro million)	11,504.7
Exports (% of GDP)	19.9	Imports (% of GDP)	38.4
Exports to EU (% of total)	53.6	Imports from EU (% of total)	56.8
Sectoral share of exports(2)	%	Sectoral share of imports(2)	%
Food and agricultural raw materials(b)	22.5	Food and agricultural raw materials	5.7
Manufactured goods(c)	42.2	Manufactured goods	24.6
Machinery and transport	17.7	Machinery and transport	20.2

[1] Purchasing Power Parities (PPPs) are currency conversion rates that both convert to a common currency and equalize the purchasing power of different currencies. In other words, they eliminate the differences in price levels between countries in the process of conversion (Permanent URL: www.oecd.org/std/ppp).

Table 2.2. Key indicators of Serbia's economy.

equipment(d)		equipment	
Chemical products(e)	7.9	Chemical products	12.7
Mineral fuels and lubricants(f)	4.6	Mineral fuels and lubricants	14.9
Other(g)	5.0	Other	22.0

Source: (1) Statistical Office of the Republic of Serbia, (2) National Bank of Serbia, Statistics, (3) World Bank, (4) Own calculations from sources above.
Note: (a) Exchange rate RDS/euro= 93.93666 and USD/euro=1.39015 applied by the Statistical Office of the Republic of Serbia, (b) Sum of 'Food', 'Beverage and Tobacco' and 'Animal and vegetable oils and fats' from Standard International Trade Classification (SITC), (c) Sum of 'Manufactured goods' and 'Miscellaneous manufactured articles' from SITC, (d) Machinery, apparatus and transport equipment from SITC, (e) Chemical products from SITC, (f) Mineral fuels and lubricants, (g) Sum of 'Crude inedible materials' and 'Commodities and transactions, n.e.s.' from SITC, (h) Indexes above 100 mean the appreciation and those below 100 the depreciation of the dinar.

2.3 Key features of the Serbian economy

The decade of the 1990s was marked by the decline of the Serbian economy. Balkan wars and international economic sanctions were main drivers of the economic decline. Since 2000 the Serbian economy has been steadily growing in terms of GDP and GDP per capita. Inflation has been significantly reduced since 2001. Increasing export activity has been one of the drivers of the economic recovery. Higher domestic consumption has driven imports significantly up (OECD, 2002). This combined with comparatively smaller growth of exports has led to a growing current account deficit. High unemployment rates remain a main problem in Serbia.

In 2009 Serbian GDP per capita was approximately 17 per cent of the EU-27 level. Table 2.3 shows economic indicators for Serbia, several neighbouring countries and the EU-27. For comparison purposes GDP is also given in PPP. Serbia's average per capita GDP in PPP is low compared to Croatia, Poland, Hungary and the EU-27, but not so far from the average levels in Romania or Bulgaria.

Table 2.3. General economic indicators for several countries and EU-27, 2009.

	Serbia (b)	Croatia (a)	Bulgaria (a)	Romania (a)	Poland (a)	Hungary (a)	EU-27 (a)
GDP (current million euro)	29,967	45,343	35,048	115,894	309,374	92,770	11,781,501
Population (million people)	7.3	4.4	7.6	21.5	38.1	10.0	498.6
GDP per capita (current euro)	4,093	10,231	4,620	5,395	8,109	9,257	23,627
GDP per capita (current euro) EU=100	17.3	43.3	19.6	22.8	34.3	39.2	100
GDP per capita, PPP (current euro)	7,175	14,377	9,977	10,271	13,599	14,611	22,576
GDP per capita, PPP (current euro) EU=100	31.8	63.7	44.2	45.5	60.2	64.7	100
Unemployment rate (% of total labour force)	17	9	7	7	8	10	9

Sources: (a) World Bank;(b) Statistical Office of the Republic of Serbia, 2010.
Note: Exchange rate USD/euro=1.39015 applied by the Statistical Office of the Republic of Serbia.

The economic decline in Serbia triggered by the global crisis in 2009 was lower than in other countries, but the impact on employment has been more dramatic. The following sections discuss the economic growth, inflation, employment, trade, balance of payments and external debt.

2.3.1 Economic growth

In the 1990s Serbia was involved in the disintegration of the Socialist Federal Republic of Yugoslavia (Slovenia 1991, Croatia 1991, Bosnia 1992-1995). From 1992 in the post-socialist transition period the country's economy was severely hit by the consequences of economic sanctions introduced by the UN (1992-1996) and the conflict in Kosovo and Metohija province (1997-1999), which ended with the NATO bombing (March-June 1999) causing significant damages to the infrastructure. The result was a severe recession in 1999 and the economy contracted by 11 per cent in one year. The period until 1999 was characterized by hyperinflation.

After the turmoil of the 1990s and the fall of the Milosevic regime in 2000, Serbia made significant progress with important democratic and economic reforms (Jusic, 2009). The restoration of the macroeconomic stability provided the basis for rapid economic growth, and since 2000 real Gross Domestic Product (GDP) has been steadily rising from 887 billion RDS constant 2002 (15.1 billion euro) to 1,352 billion RDS constant 2002 (16.6 billion euro) in 2008. However, the global economic crisis of 2008-2009 also hit Serbia, illustrated by the slowdown of economic growth already by the end of 2008 which turned into a recession in 2009. The economy shrank by 3.1 per cent in 2009 compared to the previous year. The development pattern of GDP annual growth rates oscillates from peaks of 5 per cent in 2000, 8 per cent in 2004 and 7 per cent in 2007 to valleys of 2 per cent in 2003 and -3.1 per cent in 2009 (Figure 2.1).

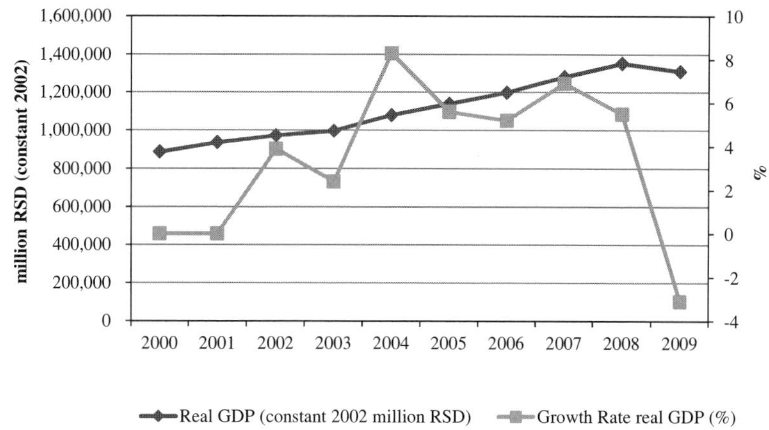

Figure 2.1. Real GDP and growth rate of real GDP in Serbia (2000-2009). Source: Statistical Office of the Republic of Serbia (2010).

In per capita terms GDP has increased from 2,324 euro (current) in 1999 to 4,093 euro in 2009. GDP per capita is growing faster than absolute GDP because of the decline in population.

GDP sectoral analysis

The Serbian economy saw rapid recovery of its economic output in 2000 after the large contraction in 1999 (see Figure 2.2). In 2000, economic growth (+5.3 per cent) was driven mainly by expansion of the industrial production and construction sector, which rose by 13.0 per cent and 16.8 per cent, respectively, while agriculture was badly affected (-12.4 per cent) by severe weather conditions (see Table 2.4). In 2001 agricultural output recovered from the droughts and floods that devastated crops and impaired livestock production during 2000, but overall agricultural production in 2001 was still less than the pre-war level and remained below until 2004. Only the sector of services was already above pre-war levels in 2003. In contrast to the agricultural decline of 1999-2000, the economic growth in the period 2000-2003 of 4 per cent (annual average) was attributed to a cautious recovery in agriculture and the growth in the service sector, while industry and construction contracted. Many industrial sectors that showed strong signs of rebound in 2000 started to contract in 2001 and continued their decline through until 2003. The year 2003 was another difficult year for the Serbian economy. Low performance of the domestic industries, along with low demand of the world economy resulted in lower than expected GDP. In 2003 agriculture was again hit severely by unfavourable weather conditions. In 2004 the Serbian economy saw growth in all sectors, with the agricultural sector growing at a remarkable rate of 19.0 per cent. In 2008 all sectors continued to grow until the global economic crisis hit the Serbian economy, impacting particularly on the sectors of industry and construction, which experienced declines in one year of 12.2 per cent and 14.3 per cent, respectively, while agriculture and services presented smoother declines. Falling foreign and domestic demand is considered the main driver behind a drop in production (-3.1 per cent) in 2009, reflecting mainly on construction and manufacturing (Stamenkovic and Petrovic, 2010).

The contraction in GDP was somewhat more moderated due to the solid performance of the agricultural and service sectors.

Table 2.4. GDP in Serbia, growth and share by activity a).

	GDP million euro current	Annual average growth real GDP (%)					GDP share (%)		
	2009	1999-2000	2000-2003	2003-2004	2004-2008	2008-2009	2009	2008	2000
Agriculture, hunting, forestry and fishing	3,139.72	-12.4	1.9	19.0	-1.3	0.7	12.1	11.7	13.9
Industry	4,762.50	13.0	-3.2	6.7	2.9	-12.2	18.3	20.3	26.7
Construction	846.64	16.8	-4.2	3.5	6.2	-14.3	3.3	3.7	4.6
Services	17,279.81	4.9	3.8	6.4	8.8	0.8	66.4	64.3	54.8
Activities - total	*26,028.67*	*4.5*	*1.4*	*8.2*	*6.0*	*-2.4*	*100.0*	*100.0*	*100.0*
FISIM	828.95	-6.0	-10.7	11.1	11.8	4.7			
GVA at basic prices	*25,199.72*	*4.9*	*1.8*	*8.1*	*5.8*	*-2.6*			
Taxes minus subsidies on products	4,767.28	8.3	19.1	9.3	5.7	-5.7			
GDP at market prices	*29,967.00*	*5.3*	*4.0*	*8.3*	*5.8*	*-3.1*			

Notes: a) 2009 are preliminary data; b) FISIM: Financial Intermediation Services Indirectly Measured; c) Exchange rate RSD/euro 93.93666.
Source: Statistical Office of Serbia (2010).

The country's output structure has changed over the last decade. The share of the industrial sector dropped significantly from about 29 per cent in 1997 to 18.3 per cent in 2009. Agriculture and construction experienced decreases in their share in total GDP of 3.4 per cent and 0.5 per cent, respectively, while the sector of services has increased its share of GDP from 51 per cent in 1997 to 66 per cent in 2009. The Serbian economy has become more reliant on services and trade.

GDP country comparison

After the recession in 1999, from 2000 to 2008 Serbia presented an average annual growth of about 5 per cent, which is two per cent points below the average growth of countries like Bulgaria and Romania for the same period but above the average growth rate in the EU-27 (2 per cent). However, in 2009 the contraction of the Serbian economy was less severe than in the neighbouring countries Bulgaria and Romania and in the EU-27.

Serbia is classified by the World Bank as an upper middle income country. For comparison purposes, Table 2.5 shows real GDP growth rates for Serbia, average GDP growth rates of the country group to which Serbia is allocated and of the high income country group in the period from 1989 to 2009. Table 2.5 shows that after the turmoil of the 1990s the Serbian growth performance was better than the averages of the upper middle income and high income country groups. GDP per capita is growing at higher rates in Serbia than in the presented country groups.

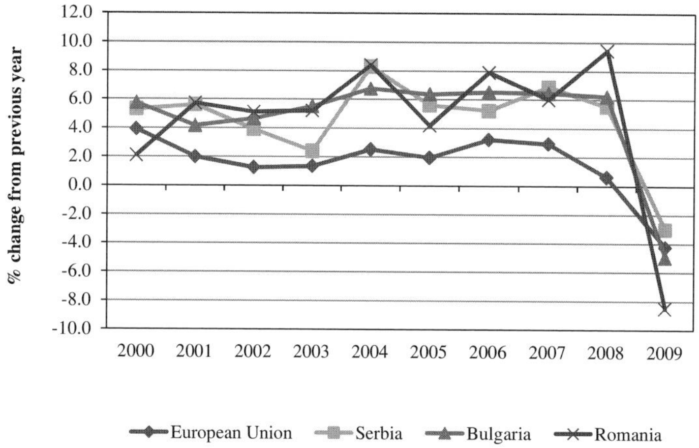

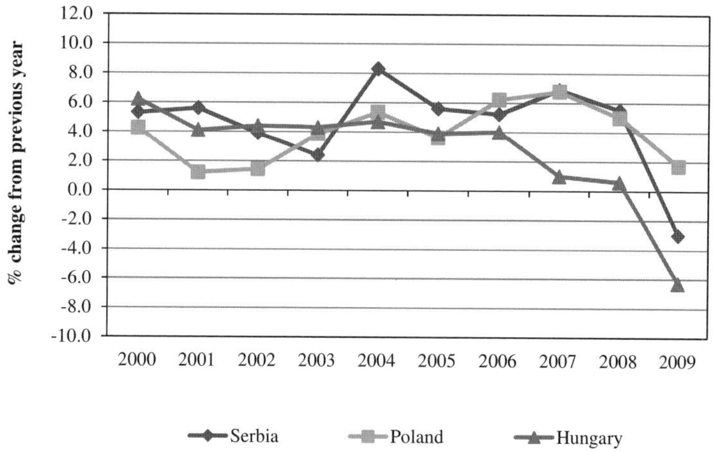

Figure 2.2. Real GDP growth rates -- Serbia, Bulgaria, Romania and EU-27 (top) and Serbia, Poland and Hungary (bottom) -- as percentage change from the previous year. Source: World Bank.

Table 2.5. Growth rates of real GDP for Serbia, upper middle and high income countries, 1989-2009 (in annual percentage).

	1990-1999	1999-2004	2004-2008	2008-2009
GDP growth in Serbia	-6.9	5.1	5.8	-3.0
GDP per capita growth in Serbia	-6.8	5.3	6.2	-2.6
GDP growth in Upper middle income	1.9	3.4	5.5	-2.6
GDP per capita growth in Upper middle income	0.7	2.4	4.6	-3.4
GDP growth in High income	2.5	2.4	2.1	-3.3
GDP per capita growth in High income	1.8	1.7	1.4	-3.9

Source: World Bank.

2.3.2 Inflation

Figure 2.3 shows changes in the Consumer Price Index (CPI) in Serbia and the Euro Area for the period 1999-2009. While changes in the CPI in the Euro Area remain for the whole period at a level of around 3 per cent, in Serbia there is a peak in 2001 (+95 per cent), after which the CPI declines considerably and since 2003 the annual change has remained at levels between +16 per cent (2005) and +8 per cent (2009). Table 2.6 shows the changes in the CPI in Serbia for the different components.

Inflation (end of the year change in percentage) for Serbia in 2009 was 6.6 per cent, moving up to 10.3 per cent in 2010 (Consumer prices) (Table 2.7).

The Agreement on Inflation Targeting between the Government of the Republic of Serbia and the National Bank of Serbia, effective as of 1 January 2009, is the formal switch of the National Bank of Serbia to inflation targeting as a monetary policy regime. The inflation targets (with a margin tolerated to divert from the target), measured as annual percentage change in the CPI, for the period 2010-2012 were established by the National Bank of Serbia (NBS) in cooperation with the Government and are the following (National Bank of Serbia 2009):

- December 2009: 8 per cent with a tolerated ±2 percentage points margin;
- December 2010: 6 per cent with a tolerated ±2 percentage points margin;
- December 2011: 4.5 per cent with a tolerated ±1.5 percentage points margin;
- December 2012: 4 per cent with a tolerated ±1.5 percentage points margin.

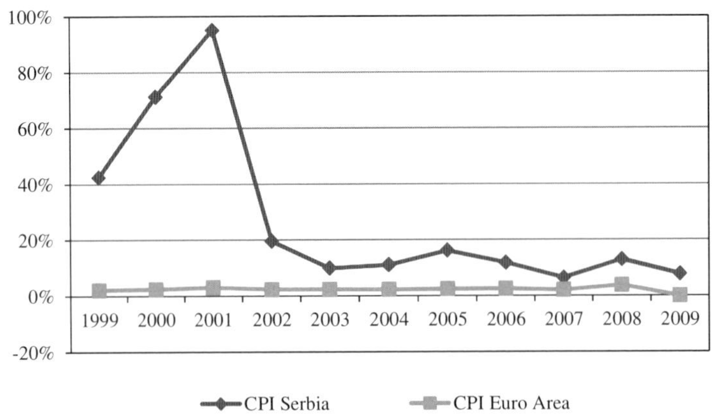

Figure 2.3. Inflation in Serbia, change in the CPI in Serbia and the Euro Area (in percentage).
Source: World Bank

Table 2.6. Changes in CPI by component (annual % growth rates).

	Total	Food	Tobacco and beverages	Clothing and footwear	Housing and household operations	Hygiene and health care	Education, culture and entertainment	Transport and PTT services
2000	71	88	63	58	32	76	61	50
2001	95	92	75	87	140	82	87	85
2002	19	9	18	16	48	8	29	26
2003	10	1	11	8	39	9	19	16
2004	11	11	10	7	25	9	10	8
2005	16	20	11	7	23	13	16	22
2006	12	10	15	11	13	11	11	13
2007	6	6	16	5	8	7	6	2
2008	13	21	10	6	5	4	7	10
2009	8	4	19	7	11	9	9	6

Source: Statistical Office of the Republic of Serbia.

Table 2.7. Consumer prices, Core inflation and Producer prices, growth rates in %.

	Consumer prices	Core Inflation	Industrial producer prices	Agricultural producer prices
2007	11.0	7.9	9.8	27.2
2008	8.6	10.8	9.0	13.5
2009	6.6	3.6	7.4	-4.8
2010	10.3	8.6	16.2	29.3

Source: National Bank of Serbia (2011).

Inflation in the last quarter of 2010 reflects the rise in inflation expectations driven by food price increases over the last several months of 2010 (see www.nbs.rs, Statistics). Inflationary pressures, to a smaller degree, also emanate from the past depreciation of the dinar and the recovery of aggregate demand. In 2010 inflation in Serbia has moved above the upper end of the target tolerance band (National Bank of Serbia, 2011).

2.3.3 Employment

Employment (total employees) in Serbia has been steadily declining since 2001 at an annual average rate of 1.5 per cent. The labour force declined in the same period from 2001 to 2009 at an average rate of 1.3 per cent, while the decline in total population of working age (+15) was less strong. Therefore, the labour force participation rate[1] dropped to 52 per

[1] Percentage of working-age (15-64) persons in an economy who are employed or are unemployed but looking for a job.

cent in 2009, which, however, was similar to the level in Poland, Romania or Croatia, and above the rate in Hungary (50 per cent) according to World Bank statistics. The employment to population (aged above 15) rate[1] in Serbia in 2008 was 44 per cent. For comparison purposes Hungary and Croatia presented rates of 45 per cent and 46 per cent, respectively, and Poland and Romania rates were each around 48 per cent. The employment to population rate in the European Union in 2008 was 50 per cent.

The International Labour Organization (ILO) defines unemployment as follows: The 'unemployed' comprise all persons above a specified age who during the reference period were (a) 'without work', (b) 'currently available for work' and (c) 'seeking work' (ILO, 1982). The definitions of employment and unemployment of Eurostat, as well as other characteristics of the Labour Force Survey (LFS), follow the definitions and recommendations of the International Labour Organization (European Commission, 2002). Since 2004 the LFS in Serbia has been conducted according to Eurostat methodology (ILO and Council of Europe, 2007). The unemployment rate in Serbia saw steady growth since 2000 and reached 14.6 per cent in 2003, the unemployment kept growing at higher rates until 2006 (21 per cent) and declined in 2008 to the level of 2002 (14 per cent). Since then a new markedly upward trend has started and the unemployment rate as of October 2010 was 19.2 per cent. These unemployment rates are surprisingly high even compared to other transition economies of the region and the EU-27 (Figure 2.4).

The labour market surveys conducted according to ILO methodology estimated the unemployment to be about half of the official registered rate, indicating the existence of a large informal employment sector. A reason for that can be found in the generous social benefits existing for the officially unemployed that encourage the informally employed population to register as unemployed (ILO and Council of Europe, 2007).

---

[1] Data from World Bank database.

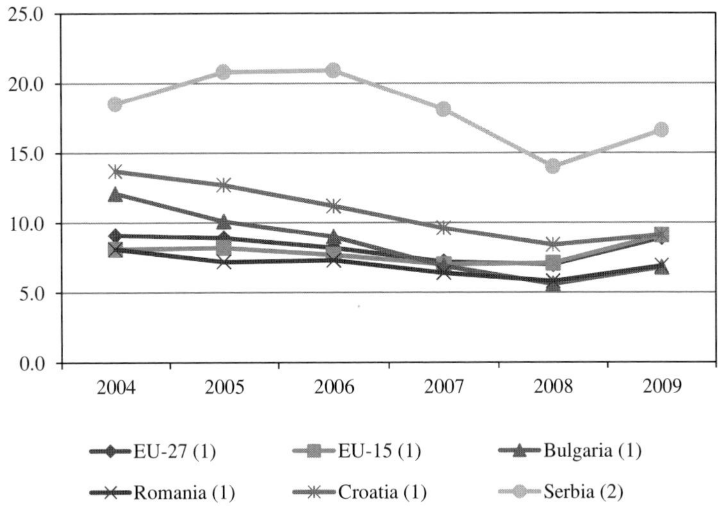

Figure 2.4. Unemployment rate in Serbia, Bulgaria, Romania, Croatia and the EU (2004-2009).
Source: (1) Eurostat (2010); (2) Statistical Office of the Republic of Serbia.

The number of employees has been declining since 2001 at an annual average rate of 1per cent, while population was declining at a lower rate of 0.3 per cent in the same period. The sectoral distribution of employment in Serbia has remained stable (see Figure 2.5). The service sector accounts for the largest portion with a 49 per cent of total employment, and industry (including construction) and agriculture account for 25 per cent and 26 per cent, respectively. For comparison purposes the share of the agricultural sector in the Euro Area (World Bank) is 4 per cent, the industry sector accounts for 27 per cent, and the service sector, similar to Serbia, accounts for the largest portion with a share of 69 per cent.

Net salaries and wages have been growing in Serbia at higher rates than GDP growth in all sectors (see Figure 2.6). Nevertheless, these data reflect only the increases in salaries and wages of official employees, and thus self-employed farmers are not included.

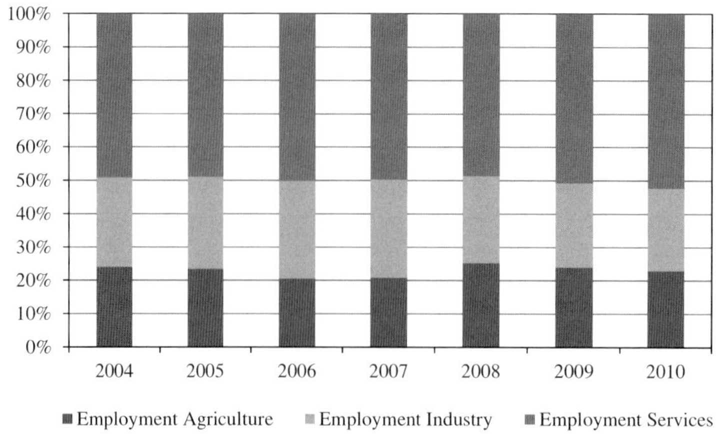

Figure 2.5. Employment by economic sector in Serbia (2004-2010).
Source: World Bank (2010).

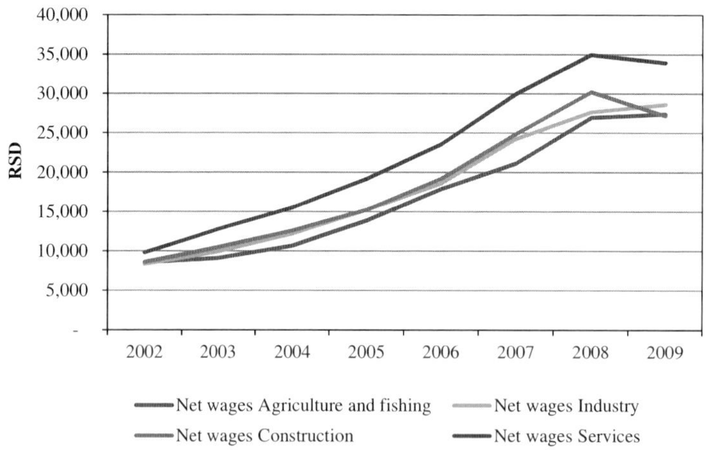

Figure 2.6. Net salaries and wages development by sector in Serbia (2002-2009).
Source: Statistical Office Republic of Serbia (2010).

2.3.4 Exchange rate

The dinar (RSD) experienced a sharp depreciating trend against the euro during the period from 2000 to 2005. After that, the exchange rate RSD/euro remained relatively stable until 2008 (see Figure 2.7). Since then depreciation pressures have continued to 2010. According to the Inflation Report of the NBS (2011), depreciation pressures in 2010 are generated by an elevated risk premium since May, markedly low capital inflows from foreign direct investments and bank and enterprise external debt repayment (see balance of payments in Table 2.10, section 2.3.6).

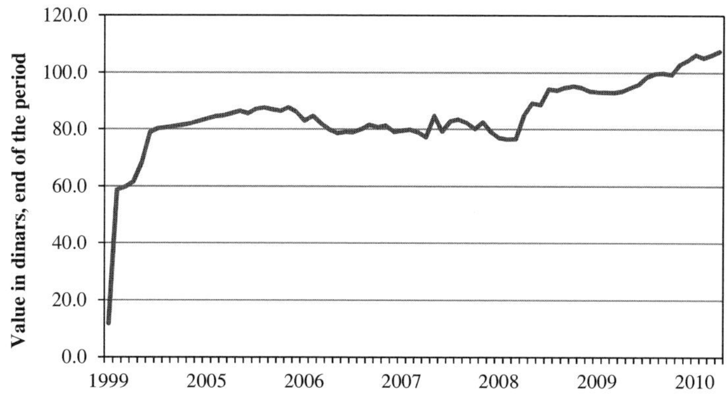

Figure 2.7. Dinar/Euro exchange rate (1999-2010).
Source: National Bank of Serbia (www.nbs.rs).

2.3.5 Trade

In 1999 exports dropped by nearly 50 per cent compared to the previous year. Thereafter, exports from Serbia have been steadily growing until 2008 at a significant annual average rate of 21.7 per cent. Exports have grown from 1,270 million euro in 1999 to 7,429 million euro in 2008. However, the global economic crisis also hit Serbia and in 2009 the value of exports by Serbia dropped by 20 per cent. Imports have followed a less regular pattern. The value of imports of Serbia have grown from 1999 to 2004 at an annual average of 26.6 per cent, and after a drop of 2 per cent in 2005, imports have followed the trend of growth until 2008 at an annual average of 24.3 per cent. Imports reached a peak of 16,478 million euro in 2008, after which a more pronounced drop took place with a decline in imports by 30 per cent from 2008 to 2009 (see Figure 2.8).

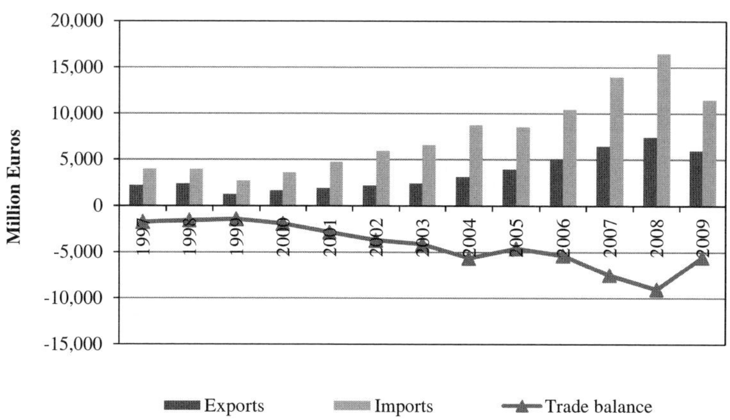

Figure 2.8. Total exports, imports and trade balance for Serbia in million euro (1997-2009).
Source: National Bank of Serbia (www.nbs.rs).

The trade deficit has been growing since 1999, except for the years 2005 and 2008 where imports experienced declines. Serbia's trade deficit decreased by 39 per cent from 2008 to 2009, from an amount of 9,049 million euro in 2008 to 5,544 in 2009 (see Figure 2.8). The trade deficit accounted for 19 per cent of GDP in 2009. Serbian trade is in deficit with all regions except for a surplus of 0.9 billion euro with the countries of CEFTA[1] (see Figure 2.9). The largest trade deficit is observed in trade with the EU, which accounts for 60 per cent of the total trade deficit of Serbia.

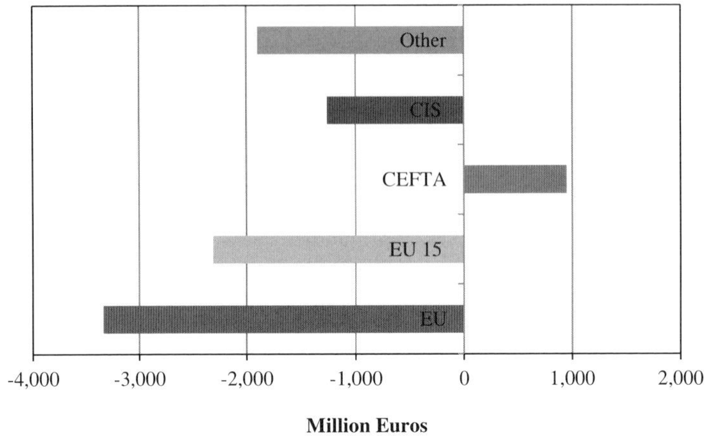

Figure 2.9. Trade balance of Serbia by regions in 2009.
Source: National Bank of Serbia (www.nbs.rs).
Note: CIS: Commonwealth of Independent States (former Soviet Republics), CEFTA: Central European Free Trade Agreement.

Most exports of Serbia (around 55 per cent) have the EU as destination, followed by CEFTA countries with a share of around 30 per cent of total exports. Most imports come from the EU (about 57 per cent), the rest originating mainly from CIS and other countries, while imports from CEFTA countries remain low (8 per cent). Figure 2.10 shows that the regional distribution of trade has remained stable.

[1] Central European Free Trade Agreement is a trade agreement between non-EU countries in Central and South-Eastern Europe. As of 2007, apart from Serbia, the parties of the CEFTA were: Albania, Bosnia and Herzegovina, Croatia, Macedonia, Moldova and Montenegro. Countries which became EU members in May 2004 are included in this group since January 2004 and at the same time excluded from CEFTA. Romania and Bulgaria are excluded from CEFTA since January 2007.

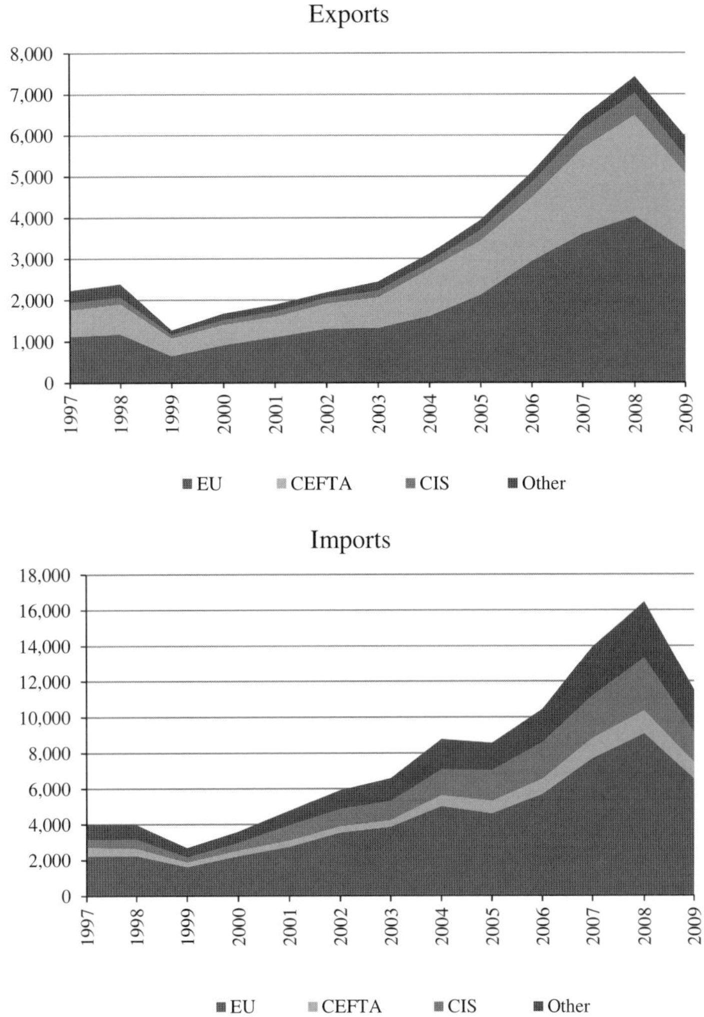

Figure 2.10. Exports and imports of Serbia by regions (1997-2009). Source: National Bank of Serbia (www.nbs.rs).

The major commodity group of exports (see Table 2.8) are 'Manufactured goods' with a share of 42 per cent in 2009. Exports of 'Food and agricultural raw materials' account for the second largest share of total exports in 2009 (22 per cent). The growth rates of the exports of these two commodity groups have been decreasing since 1999. However, while the group of 'Manufactured goods' experienced negative growth rates in the world economic crisis period 2008-2009, 'Food and agricultural raw materials' kept growing at a rate of 4 per cent. Exports of machinery and transport equipment have grown at a significant growth rate of 76 per cent between 2005 and 2008, moving from 393 million euro in 2005 to 1,294 million euro in 2008. They account for a share of 18 per cent of total exports.

Table 2.8. Exports of Serbia by SITC product classification.

	Thousand euro	Average growth rates (%)			2009 Share (%)	2008 Share (%)
	2009	2001-2005	2005-2008	2008-2009		
Food and agricultural raw materials (SITC 0+1+4)	1340	26	15	4	22	17
Manufactured goods (SITC 6+8)	2517	20	21	-38	42	47
Machinery and transport equipment (SITC 7)	1056	10	49	-23	18	17
Chemical products (SITC 5)	473	32	20	-59	8	10
Mineral fuels and lubricants(SITC 3)	276	27	20	8	5	3
Other (SITC 2+9)	299	5	27	-25	5	5
Total	5961	20	24	-25	100	100

Source: National Bank of Serbia.

The major commodity group for imports in 2009 are the 'Manufactured goods' with a 25 per cent of total imports followed by group of 'Other' (Commodities and transactions not classified elsewhere), which experience a significant growth from 2005 to 2008. 'Food and agricultural raw materials' account for 6 per cent of total imports and different to exports, imports of food declined during the world economic crisis period 2008-2009. Growth rates have been declining by all groups except for the group of 'Other', which moved from 410 million euro in 2005 to 2,529 million euro in 2009, with a growth rate between 2005 and 2008 of 288 per cent.

Table 2.9. Imports of Serbia by SITC product classification.

	Thousand euro	Average growth rates (%)			2009 Share (%)	2008 Share (%)
	2009	2001-2005	2005-2008	2008-2009		
Food and agricultural raw materials (SITC 0+1+4)		6	4	-7	6	4
Manufactured goods (SITC 6+8)	2,835	23	15	-36	25	23
Machinery and transport equipment (SITC 7)	2,321	22	14	-39	20	20
Chemical products (SITC 5)	1,456	13	14	-21	13	11
Mineral fuels and lubricants(SITC 3)	1,712	15	22	-74	15	18
Other (SITC 2+9)	2,529	-6	113	-56	22	24
Total	11,505	16	24	-43	100	100

Source: National Bank of Serbia.

The value of Serbian exports currently represents 20 per cent of GDP. One of the main objectives of the export-oriented growth projected for 2020 is to achieve an average annual growth of exports of 13.5 per cent, reaching a share of 65 per cent in GDP in 2020. This implies an increase in exports from 5,961million euro in 2009 to 34,200 million euro in 2020 (Stamenkovic and Petrovic, 2010).`

2.3.6 Current account and external debt

Serbia's current account deficit has been increasing significantly from 2005 until 2008. Since then, mainly due to the effects of the global economic crisis it has decreased.

Serbia's current account deficit was 7.054 million euro in 2008 (21 per cent of GDP), increasing by almost 62 per cent (annual average) since 2005, when the deficit amounted to 1,778 million euro and had experienced a 32 per cent decrease compared to 2004. As shown in Figure 2.11, the current account balance has deteriorated sharply since 2005 to 2008, which, however, was mainly a result of a significant increase in imports accompanying the strong economic growth (average annual GDP growth of 6.2 per cent in the period 2005-2008). At the same time, in 2006 the economy also saw a remarkable increase in current transfer outflows, which rose to an estimated 1,051 million euro in 2006, thus putting more pressure on the current account deficit. However, this factor disappeared in the following years (see Table 2.10). The global economic crisis that has also hit the Serbian economy brought the current account deficit to 2,084 million euro in 2009 and 1,992 in October 2010. The improvement of the current account deficit is mainly due to the sharp decrease in imports, which in October 2010 were below the levels of 2006, while exports saw less dramatic declines.

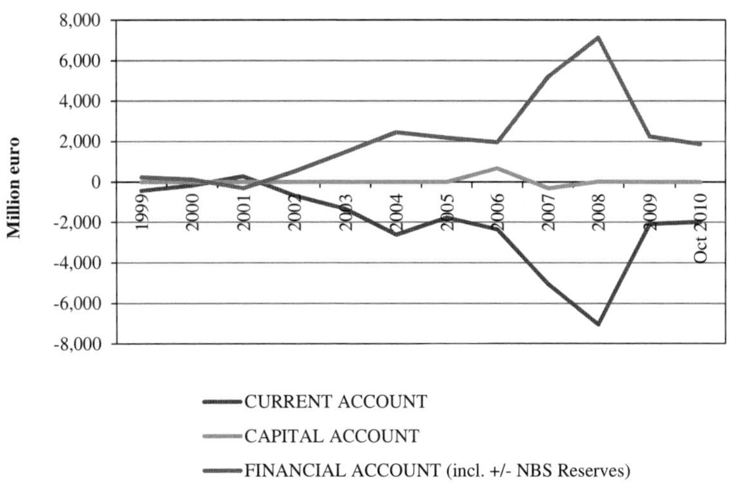

Figure 2.11. Balance of payments of Serbia 1999-2010 a) (October).
a) 2010, estimated data 3rd trimester.
Source: National Bank of Serbia.

Table 2.10. Balance of payments of Serbia.

in million euro	2004	2005	2006	2007	2008	2009	Oct 2010
Current account	*-2,620*	*-1,778*	*-2,356*	*-5,053*	*-7,054*	*-2,084*	*-1,992*
in per cent GDP	-13.8	-8.8	-10.1	-17.6	-21.1	-7.0	-
From which							
Export of goods f.o.b.	3,286	4,009	5,109	6,383	7,416	5,978	5,960
Import of goods f.o.b.	-8,487	-8,287	-10,090	-13,451	-15,917	-11,096	-9,870
Services	133	-5	-41	-261	-185	19	13
Current transfer expenditure	-479	-674	-1,051	-229	-275	-244	-220
Capital account	-	-	*670*	*-314*	*13*	*2*	*1*
Financial account (incl. +/- NBS Reserves)	*2,448*	*2,181*	*1,957*	*5,176*	*7,133*	*2,237*	*1,871*
From which							
Foreign direct investment-net	774	1,250	3,323	2,513	2,018	1,439	775
Portfolio investment-net	-	-	355	678	-91	-55	91
Loans (liabilities)	1,252	1,805	2,292	3,402	3,529	1,387	424
NBS reserves assets(-increase)	-343	-1,647	-4,269	-742	1,687	-2,363	1,080
Errors and omissions – net	172	-403	-270	191	-92	-155	121
Overall balance	*343*	*1,647*	*4,269*	*742*	*-1,687*	*2,363*	*-1,080*

Source: National Bank of Serbia.

Not only the size but also the sources of financing are important for the sustainability of the current account deficit. The most sustainable source is foreign direct investment. Serbia's large current account deficit has been comfortably balanced by significant foreign direct investments in 2005 and 2006. However, since then this source was declining and the current account deficit needed to be compensated with significant debt-creating capital inflows since 2007. In 2008 the last source dropped considerably and NBS reserves were reduced by 1,687 million euro. In 2009 the overall balance is positive because of the significant decline in the current account deficit (imports driven) and in 2010 (October data) despite current account deficit stays at levels of 2009 the overall balance is negative by 1,080 million euro due to the important decrease in both foreign direct investments and loans (see Figure 2.12).

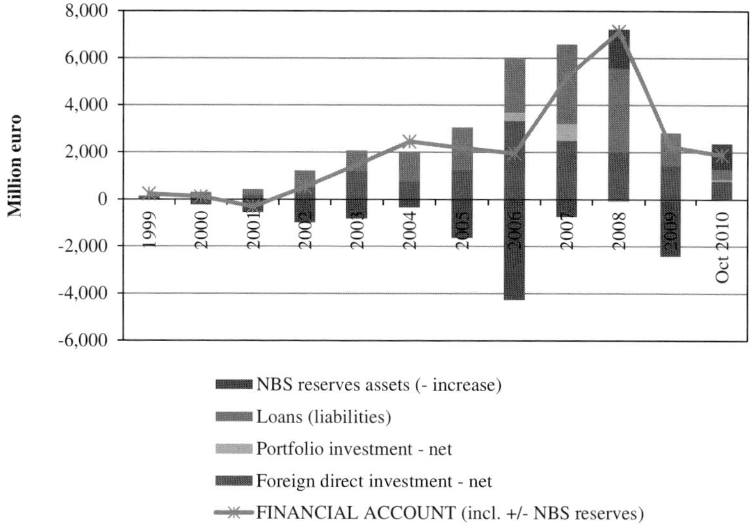

Figure 2.12. Structure of the financial account of Serbia (1999-2010).
Source: National Bank of Serbia.

The total external debt in Serbia reached 75 per cent of GDP at the end of 2009, which represents a significantly higher level of indebtedness than the average of the upper middle income countries group in which Serbia is classified (according to World Bank classification). The public sector accounts for 35 per cent of the external debt and consequently the private sector accounts for the remaining 65 per cent.

Figure 2.13 shows the Serbia's external debt by creditor and illustrates how since 2004 the external debt coming from creditors other than governments and international financial organizations is growing significantly.

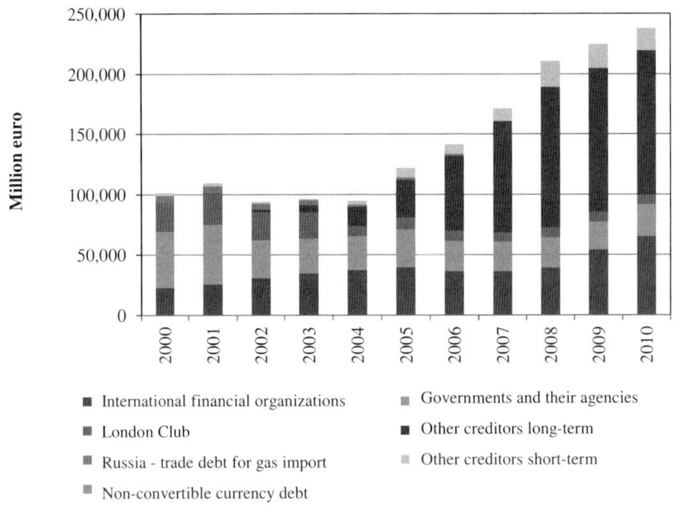

Figure 2.13. Serbia's external debt by creditor.
Source: National Bank of Serbia.

2.4 Measuring the size of the economy

The size of the Serbian economy may be reported in real or nominal GDP, denominated in RSD or in another currency. The GDP may be converted into a foreign currency at market exchange rates or at rates reflecting the standardized purchasing power (Purchasing Power Parity rates, PPP). The size of two economies may be indifferently compared at a certain point by using nominal or real units; however, the use of PPP is recommended when comparing living standards between economies. When measuring the GDP growth of a country, real units are recommended and the choice of deflator to remove the effect of price inflation becomes relevant.

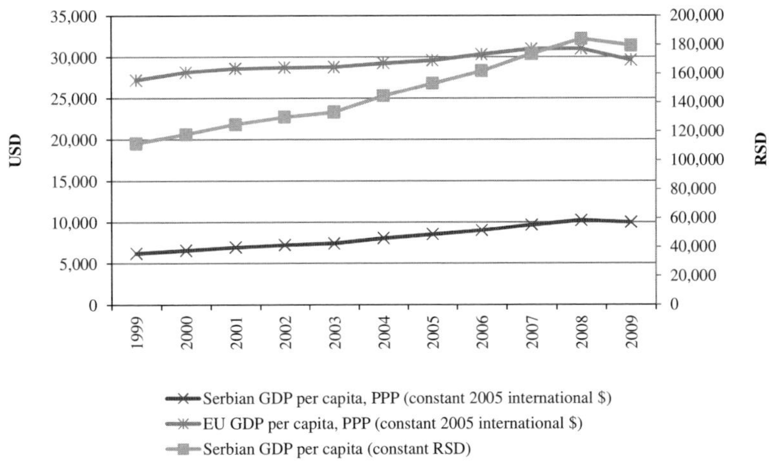

Figure 2.14. Trends in per capita GDP in Serbia and the EU (1999-2009). Source: World Bank.

Figure 2.14 shows GDP per capita developments for Serbia and the EU, presenting GDP per capita in PPP for Serbia and the EU (both in USD) and real GDP for Serbia in the local currency (RSD). The size of the Serbian economy measured is larger when using PPP rather than USD. The gap between the two series in USD has been widening since 1999 reflecting the continuous depreciation of RSD against USD (and also the euro). The EU and Serbia measured in PPP has remained quite stable in the period from 1999 to 2009.

Serbia's future contribution to the EU budget would depend on its GDP measured in euro. The gap between per capita GDP in Serbian regions and average EU per capita GDP, both measured in euro, is relevant for the eligibility for structural funds. Conversion to euro for this purpose would be normally performed at market exchange rates. This issue will be further considered in section 11.2.

2.5 Income distribution

This section provides an overview of the income distribution in Serbia. Section 2.5.1 looks at the distribution of income over households, and compares household income distribution over time (in terms of GINI coefficient), between rural and urban areas, and with the income distribution of neighbouring countries. Furthermore, this section provides an analysis of poverty in Serbia. Section 2.5.2 considers regional income distribution at the NUTS 2 level. Serbia is compared to selected EU Member States in terms of regional income disparities.

2.5.1 Developments in the household income distribution

Table 2.11 provides data on the household income distribution in Serbia and compares it with the countries Croatia and Romania. In Serbia the richest 20 per cent of households received around 38 per cent of total disposable income in 2008. In Romania and Croatia the richest 20 per cent received similar percentages of total income (data of 2007 and 2005, respectively). The Serbian situation is comparable to the EU-15 countries; taking Germany and Spain as examples, the richest 20 per cent had income percentages of 37 per cent and 42 per cent, respectively.

Table 2.11. Household income distribution in Serbia and neighbouring countries.

Percentage of Households	Serbia Income share (%)	Romania Income share (%)	Croatia Income share (%)
	2008	2007	2005
Income share held by highest 20%	37.5	40.3	37.9
Income share held by fourth 20%	22.5	22.3	22.7
Income share held by third 20%	17.4	16.8	17.3
Income share held by second 20%	13.6	12.7	13.3
Income share held by lowest 20%	9.1	7.9	8.8

Source: World Bank.

The Gini index is an inequality measure, comparing the actual distribution with a hypothetical situation where everybody gets an equal share. The Gini index varies from zero to one, zero representing perfect equality and one representing the maximum possible degree of inequality (one household receives everything). Table 2.12 provides information on the Gini index of Serbia, urban and rural from 2002 to 2008, with the index presented on a scale of zero to 100. The Gini index of Serbia has slightly diminished in this period. Inequality in urban areas measured by the Gini coefficient was slightly lower compared to rural areas in 2002, while five years later the inequality in urban areas was higher compared to rural areas. According to Gini Index comparisons, EU-27's income inequality is higher than in Serbia.

Table 2.12. Gini Index in Serbia 2002-2008.

	2002 a)	2004 a)	2007 a)	2008 b)
Gini index Serbia	29.29	29.7	29.7	28.18
Urban	28.38	-	29.53	-
Rural	29.72	-	27.58	-
Gini index EU 27			30.6	30.7

Source: a) Krstic (2008); b) World Bank.

The next table (2.13) provides information on the GDP per capita and Gini index of Serbia and its neighbouring countries.

Table 2.13. GDP per capita and Gini index for Serbia and selected countries, 2004.

	GDP per capita (USD) 2004	Gini Index
Slovenia	16,115	28.4
Hungary	9,962	26.9
Croatia	7,724	29.0
Romania	3,374	31.0
Bulgaria	3,109	29.7
Serbia	2,835	29.7
Albania	2,439	28.2
Bosnia and Herzegovina	2,183	26.2

Source: Statistical Office Republic of Serbia (2010).

A main disadvantage of the Gini index is that it gives the same values both in cases when the inequality is greater in the lower deciles (poor parts of the population) and in the higher deciles (wealthiest part of the population) (Jovičić and Milojević, 2010). Poverty intensity is usually measured by poverty depth or poverty gap. It measures how far off households are from the poverty line. Poverty severity takes into account not only the distance separating the poor from the poverty line, but also the inequality among the poor.

Poverty in Serbia considerably decreased in the 2002-2007 period (Table 2.14), in both urban and rural areas. In 2002, 14 per cent (approximately 1 million people) were poor, compared to 6.6 per cent or some 0.5 million in 2007. The depth and severity of poverty, which reflect the poverty distribution, also decreased in this period (see section 5.6 for more detailed analysis of rural poverty, with more up-to-date data).

Table 2.14. Key poverty indicators in Serbia by settlement type, 2002-2007.

	Poverty index (% of population)			**Poverty depth (%)**			**Poverty severity (%)**		
	2002	2007	Change	2002	2007	Change	2002	2007	Change
Urban	11.2	4.3	-6.8	2.1	0.8	-1.3	0.6	0.3	-0.4
Rural	17.7	9.8	-8.0	4.2	2.0	-2.2	1.5	0.6	-0.9
Total	*14.0*	*6.6*	*-7.4*	*3.0*	*1.3*	*-1.7*	*1.0*	*0.4*	*-0.6*

Source: Krstic (2008).

2.5.2 Regional income distribution

The regions of Belgrade and Vojvodina, which compose the Serbia-North territory (NUTS 1 level), account for 66 per cent of total GDP, while Serbia-South accounts for the remaining 34 per cent. In per capita terms, the GDP in the Belgrade region is 2 times that found in Vojvodina, 2.5 times that found in Šumadija and Western Serbia and 3 times that found in Southern and Eastern Serbia (see Table 2.15). The regional income inequalities arise mainly because of the differences in types of economic activities and the related differences in productivity across regions. For instance, poorer regions generally have a bigger share of their resources employed in agriculture, where productivity is usually lower. An analysis of the degree of rurality in Serbia (Gligorijević and Stepić, 2010) shows that Serbia is highly polarized into the urban region of Belgrade, at one end of the spectrum, while the remaining territory consists of rural areas.

Serbia shows slightly higher regional income disparities when compared to selected EU-15 and NMS countries. The average ratio of GDP per capita between the richest and the poorest NUTS 2 regions averages 1.8 in the selected countries, whereas it is 2.8 in Serbia (Table 2.16).

Table 2.15. Regional GDP, 2009 preliminary results.[1]

Region (NUTS 2)	GDP (million RSD)	Share (%)	GDP per capita ('000 RSD)	Level index RS=100
Republic of Serbia	2,815,000.00	100	385	100
Belgrade region	1,124,565.00	39.9	690	179.4
Vojvodina region	720,301.00	25.6	366	95.2
Region of Šumadija and Western Serbia	563,734.00	20.0	275	71.4
Region of Southern and Eastern Serbia	406,400.00	14.4	243	63.3

Source: Statistical Office of the Republic of Serbia (2010).

Table 2.16. Regional disparities in terms of GDP per capita ratio at NUTS Level 2, 2007.

Country	Region with the highest GDP per capita (euro)		Region with the lowest GDP per capita (euro)		Disparity ratio
Germany	Hamburg	49,000	Thueringen	21,200	2.31
Netherlands	Groningen	43,300	Flevoland	28,200	1.54
Italy	Bolzano	33,800	Calabria	16,600	2.04
Bulgaria	Jugozapaden	6,200	Yuzhen tsentralen	2,700	2.30
Romania	Vest	6,700	Macroregiunea unu	5,700	1.18
Serbia	Belgrade region	8,708	Southern and Eastern Serbia	3,067	2.84

Source: EUROSTAT and Statistical Office of the Republic of Serbia (2010).

2.6 Summary

Serbia is a country located in the central south-eastern part of Europe with a population of 7.3 million people and a per capita income of 4,100 euro in 2009. Population is declining at an annual average rate of 0.3 per cent.

Economic growth

The decade of the 1990s in Serbia was marked by the economic decline driven by the Balkan wars and the international economic sanctions. Since 2000 the Serbia economy has steadily been growing in terms of GDP (4.4 annual average) and GDP per capita (4.7 per cent annual average). Increasing export activity has been one of the drivers of the economic recovery. The global economic crisis also hit the Serbian economy and GDP has decreased in 2009 by 3 per cent. The GDP growth of Serbia is about 2 per cent below the average growth of countries like Bulgaria and Romania but around 2 per cent above the EU-27 and above the average growth rate of the upper income countries group in which Serbia is

[1] The calculations of regional GDP, as a part of the system of national accounts of the Republic of Serbia, were established pursuant to the EU principles and methodology, i.e. the European system of national accounts (ESA 95) and the Nomenclature of territorial units for statistics of the Republic of Serbia (NSTJ), which has been completely harmonized with the European Nomenclature of territorial units for statistics (NUTS). The calculations of regional GDP, in accordance with the European legislation, were carried out at the NSTJ level 2 (region) and they anticipate the following territorial units for statistics: Belgrade region, Vojvodina region, Region of Šumadija and Western Serbia, Region of Southern and Eastern Serbia, Region of Kosovo and Metohija. For this region, no reliable statistical data are at our disposal and it was not covered by the calculations.

classified. In terms of GDP, the contraction of the economy in 2009 was less severe than in the countries mentioned. The country's output structure has changed over the last decade. The share of the industrial sector has dropped significantly from 29 per cent in 1997 to 18 per cent in 2009. Agriculture and construction have experienced more moderate decreases in their share, while the services sector has increased its share considerably from 51 per cent in 1997 to 66 per cent in 2009.

Inflation

While in the period 2000-2009 CPI changes in the European Union remain at a level around 3 per cent, in Serbia there was a peak in 2001 (+95 per cent) after which the CPI declined to 8 per cent in 2009. In 2010 inflation had a rebound and moved to the upper end of the target tolerance band established by the National Bank of Serbia in cooperation with the Government.

Employment

Unemployment rates are high even compared to other transition economies of the region. The sectoral distribution of employment remains relatively stable. The services sector accounts for the largest share (49 per cent) and industry and agriculture account for 25 per cent and 26 per cent, respectively, in 2008.

Trade

Exports and imports have been steadily up from 1999 at average rates of 22 per cent and 25 per cent, respectively, until 2008. Exports dropped by 20 per cent in 2009 and the drop in imports was even more pronounced (30 per cent) driven by the reduction of consumption. The trade deficit has been growing in most years since 2000, accounting for 19 per cent of GDP in 2009. The trade of Serbia is in deficit with all regions except for the countries of CEFTA. Exports of 'food and agricultural raw materials' account for 22 per cent of Serbia's total exports, while on the import side food and agricultural imports represented only 6 per cent in 2009.

Current account and external debt

Driven by the trade deficit, the current account deficit has been increasing significantly from 2002 until 2008. Since 2008 the consequences of the global economic crisis have contributed to reduce the deficit. Serbia's current account deficit had been comfortably balanced by significant foreign investment. Since 2007 this source has been declining, and a current account deficit needed to be compensated with debt-creating capital inflows. Total external debt in Serbia reached 75 per cent of GDP at the end of 2009, reflecting a considerably higher level of indebtedness than the average of upper middle income countries. Since 2004 Serbia's external debt has been mainly financed by creditors other than governments and international financial organizations.

Income distribution

In Serbia the richest 20 per cent of households received around 38 per cent of total disposable income in 2008. In Romania and Croatia this percentage was 40 per cent and 38 per cent, respectively; in the EU-15 countries, taking Germany and Spain as examples, the percentages varied between 37 per cent and 42 per cent, respectively. The inequality in urban areas measured by the GINI index was higher in urban than in rural areas in 2007. Poverty in Serbia decreased in the period 2002-2007 in both urban and rural areas. In 2002 about 14 per cent or 1 million people were poor, while in 2007 about 7 per cent or 0.5 million people were poor.

The regions of Belgrade and Vojvodina (NUTS 1) account for 66 per cent of total GDP. In per capita terms the GDP in Belgrade regions is 2 times more than in Vojvodina, 2.5 times more than in Sumadija and Western Serbia and 3 times more than in Southern and Eastern Serbia. The regional income inequalities arise mainly because of the differences in the type of economic activities. Poorer regions generally have a bigger share of their resources devoted to agriculture.

Literature and websites

EC, *Commission regulation (EC) No 1897/2000 of 7 September 2000 implementing Council Regulation (EC) No 577/98 on the organization of a labour force sample survey in the Community concerning the operational definition of unemployment.* European Commission, Brussels, 2002.

Eurostat, *Statistic Database*, Brussels/Luxembourg. Available at http://epp.eurostat.ec.europa.eu.

FAO, *FAOSTAT*. Rome. Available at http://www.faostat.fao.org.

Gligorijević, V. and M. Stepić, *Rural regions in Serbia - population structure and change*. 2010. Available at: www.demobalk.org/conferences/fr/BUDVA_10_5-2010/S_3/Gligorijevic_S3.pdf

ILO, *Resolution concerning statistics of the economically active population, employment, unemployment and underemployment, adopted by the Thirteenth International Conference of Labour Statisticians.* ILO, Geneva, 1982.

ILO and Council of Europe, *Employment policy review Serbia.* ILO, Geneva, 2007. Available at: www.ilo.org/public/libdoc/igo/2007/410625.pdf

Jovičić, M. and T. Milojević, *Poverty and inequality changes in Serbia as the result of global instability.* The 11th Bi-Annual Conference of the European Association for Comparative Economic Studies. August 26-28, 2010, Tartu, Estonia: Comparing Responses to Global Instability, 2010.

Jusic, A., *Integration perspectives and synergic effects of European transformation in the countries targeted by EU enlargement and neighbourhood policies: case study of Serbia.* Publications of the Center for EU Enlargement Studies - Central European University, Budapest. 2009. Available at: www.ssrn.com/abstract=1352522

Krstic, G., *Poverty profile in Serbia in the period from 2002-2007.* Living Standards Measurements Study-Serbia 2002-2007 (LSMS). Statistical Office of the Republic of Serbia, Belgrade, 2008.

National Bank of Serbia, *Inflation Report.* Available at: www.nbs.rs/export/internet/english/90/90_5/inflation_report_february_2011.pdf. Accessed 20 April 2011.

National Bank of Serbia, *Memorandum of the National Bank of Serbia on setting inflation targets for the period 2010-2012.* 2009. Available at: www.nbs.rs/export/internet/english/30/memorandum_ciljevi_2010_2012_eng.pdf

National Bank of Serbia, *Statistics.* Available at: www.nbs.rs/export/internet/english/80/index.html. Accessed 21 April 2011.

OECD, *OECD Economic Surveys 2001-2002. Federal Republic of Yugoslavia, Economic Assessment.* OECD, Paris, 2002.

SORS (Statistical Office of the Republic of Serbia), *Regional GDP of the Republic of Serbia - preliminary results.* SORS, Belgrade, 2010.

SORS (Statistical Office of the Republic of Serbia), *Statistical Yearbook 2010.* Available at: http://webrzs.stat.gov.rs/axd/en/

Stamenkovic, S. and P. Petrovic, *Serbian post-crisis economic growth and development model 2011-2020.* USAID, The Economic Institute MAT and The Faculty of Economy-FREN, Belgrade, 2010.

World Bank. Data from http://data.worldbank.org/

Annex to Chapter 2 Political developments over the last century

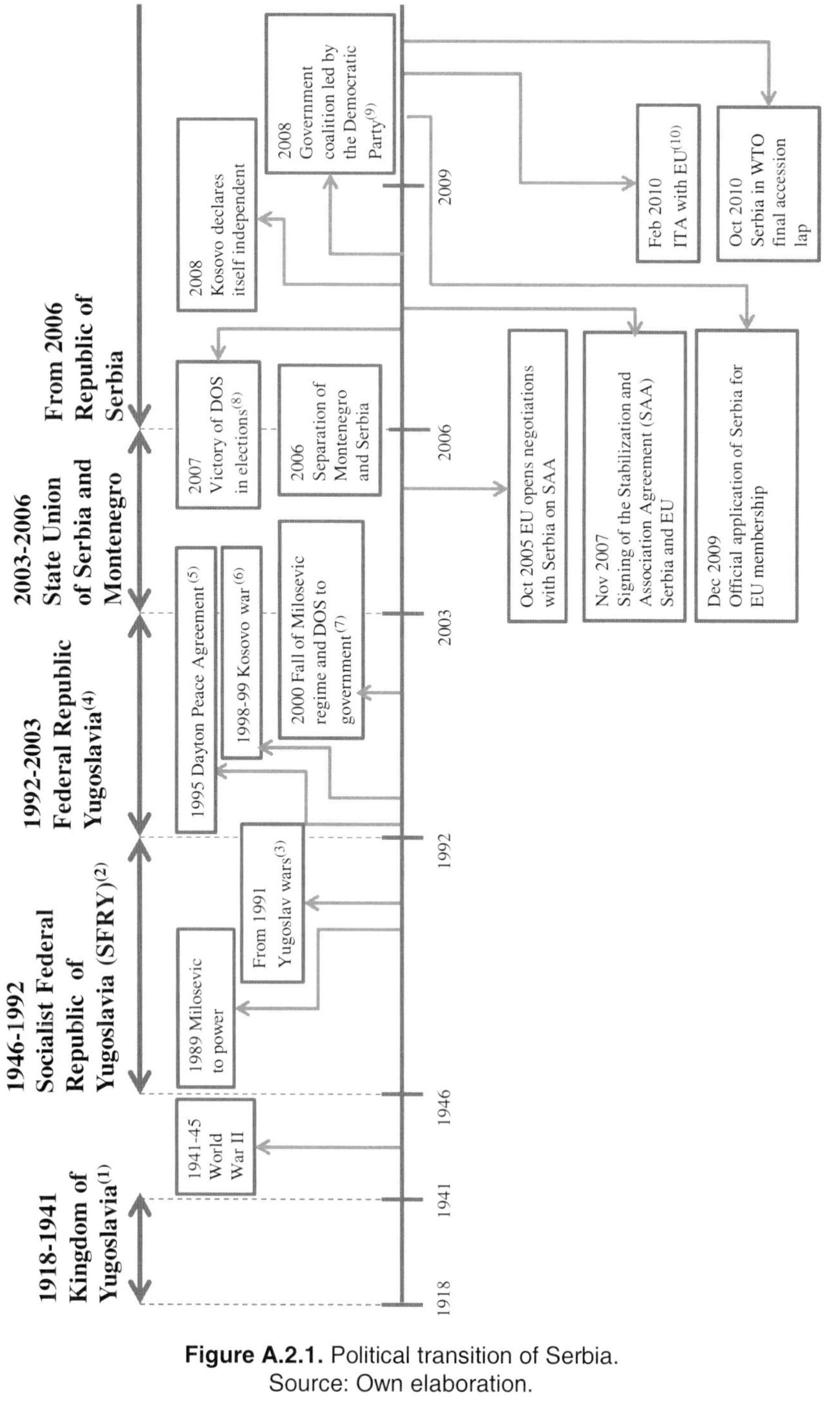

Figure A.2.1. Political transition of Serbia.
Source: Own elaboration.

Note:
1) Formerly named Kingdom of Serbs, Croats and Slovenes. Consists of the States of Slovenes, Croats and Serbs and the Kingdom of Serbia to which the Kingdom of Montenegro was annexed in November 1918.
2) Six Socialist Republics (SR) and two Socialist Autonomous Provinces (SAP): SR Bosnia and Herzegovina, SR Croatia, SR Macedonia, SR Montenegro, SR Slovenia and SR Serbia (including the autonomous provinces of SAP Vojvodina and SAP Kosovo which after 1974 were largely equal to the other members of the federation.
3) Followed the secession of most of the country's constituent entities.
4) April 1992 Communist Yugoslavia ceases to exist.
5) End of the Bosnian War that internally partitioned Bosnia and Herzegovina largely into a Serb republic and a Bosnia-Croat federation.
6) Since then, Kosovo is under UN administration.
7) DOS: Democratic Opposition of Serbia.
8) In the general elections of early 2007, victory of the reformist Democratic Party (DS).
9) A cabinet assembled by the Prime Minister Mirko Cvetković, supported by coalition 'For a European Serbia' (DS, G17+, SPO, LSV, SDP), coalition SPS/PUPS/JS and minority parties SVM, SDPS, BDSS and SLPS. Opposition parties are SRS, DSS, NS, SNS, PzDD and LDP.
10) Full implementation of the Interim Trade Agreement (ITA) with the EU.

Figure A2.1 shows the political developments of Serbia in the time line since the beginning of the 20th Century until early 2010.

The decade of the 1990s in Serbia has been marked by wars and international economic sanctions that have been the main cause for the economic decline in this decade.

In 2000, a change of government at the federal level, and then in Serbia, marked a turning point. The new government implemented stabilization measures and initiated a market reform development programme. With the ambitious programme of economic reform of the new government, Serbia was committing itself to reversing the trends of the previous decade and integrating the economy with the EU. In less than two years trade and prices have been liberalized; new fiscal and monetary policies have contributed to reduce inflationary pressure, stabilized the exchange rate and improved expectations (OECD, 2002). The programme of reforms, designed together with international economic institutions, promoted the structural changes needed to sustain growth into the future.

Since 2000 the Federal Republic of Yugoslavia has started the reintegration with the international community (renewal of membership in the International Monetary Fund, World Bank and European Bank for Reconstruction and Development). Serbia is currently in the final lap of accession to the WTO.

European integration is a stated priority of the current Government of Serbia and major progress has been made towards EU accession with the signing of the Stabilization and Association Agreement (SAA) with the EU in 2007 and with full implementation of the Interim Trade Agreement (ITA) with the EU in February 2010. In December 2009 Serbia officially applied for EU membership.

Chapter 3

3 The institutional framework of Serbia and Serbian agriculture

3.1 Introduction

This chapter describes the current institutional framework of the Serbian agricultural sector, beginning with a brief account of the key historical developments that have influenced this framework. Specifically, we examine institutional developments that have implications for the alignment of Serbian agricultural institutions with those of the EU, with the aim of providing insights relevant to Serbia's possible accession to the EU and to expectations of future developments.

3.2 Historical overview of institutional development

The most significant changes and major shifts in the socio-economic and social system of Serbia are listed in this review. For the period from the end of WWI to 1990 the term Yugoslavia is used, as the institutional arrangements relevant to this analysis, was universal for the country as a whole.

Economically exhausted, with great human and material losses caused by the Balkan and the First World War, Yugoslavia was during the first half of the 20th century one of Europe's economically least developed and poorest countries. In that period Yugoslavia's economy was dominated by small, self-subsistence peasant farms. Most of the country's few industries were owned by foreigners.

Post World War II development of Yugoslavia has been radical and dynamic. During the first five years Yugoslavia followed Stalin's model of a centrally planned economy, including the Soviet-style agricultural collectivization. Political disagreement between the Yugoslav government and Stalin and the Cominform in 1948 pushed the country into an entrenched economic isolation. This change meant that trade with socialist countries fell, and the country had to turn to Western European countries in order to obtain equipment, fuel, and raw materials for its intensive industrial development. Throughout the 1950s, USA and West European credits and grants were vital in sustaining industrial growth.

At the beginning of the 1950s the government formally abandoned state ownership and central planning in favour of programmes known as social ownership and workers' self-management. Under the principle of self-management, the workers theoretically ran the factories and were free to make management and marketing decisions. Even with such a façade, the state still controlled the economy, having the power to assign enterprise directors and to allocate funds to inefficient enterprises. Agriculture remained with primitive farming methods while farmers had no access to bank loans or other sources of financing. These factors limited food production, forcing Yugoslavia to import grain and other foodstuffs. Millions of peasants migrated to cities and tried to find jobs in industries.[1]

In 1965 the government introduced economic reforms and established a new programme that was promoted as market socialism. Under this programme, the government left control of investment funds to enterprises and to a reformed banking system. The country's economy made significant progress. Unemployment was low and the education level of the working force gradually increased. Social equity and equality of all federal republics, minorities, genders, and the different layers of society were the basic principles promoted by the leading party. Due to Yugoslavia's neutrality and important role in the Non-Aligned Movement, Yugoslav companies exported to both Western and Eastern markets and carried out the construction of numerous infrastructure and industrial projects in Africa, Europe and Asia. The fact that Yugoslavs were allowed to emigrate freely in the 1960s caused many to find work in Western Europe. The emigration was mainly caused by a rural exodus and abandonment of agriculture. The earnings they sent home became important to the Yugoslav economy, contributing to a boom in private house building and SMEs. By 1970 about 25 per cent of the economically active rural population held jobs outside agriculture, and about half of Yugoslavia's rural population lived in 'mixed' households (with at least one member who held an industrial job).

In 1974 the government ratified a new Constitution which created a complex system of balance of power. The Constitution introduced the consensus among republics and provinces in decision-making. The Constitution tried to refine the balance between economic and ethnic diversity on the one hand, and the communist ideal of social unity on the other. The economic and investment priorities changed to favour consumer goods more than heavy industries.

The agricultural sector was inefficient until the 1970s. The food distribution system was dysfunctional, farm incomes were low, and agricultural prices were under strict control. Most investment was earmarked for the social sector, and vertically integrated state farms had higher yields than private farms. The longstanding imbalance of agricultural investment and the importance of the private sector were recognized in the 1980s by the federal government's Green Plan, which devoted attention and funding to agriculture. This was the first governmental initiative in favour of the private sector and helped to increase both productivity and living standards. Peasant farmers gained access to credits, enabling them to buy farm machinery to expand production and to end the country's dependence on food imports. The Green Plan provided premiums for the production of wheat, sugar beets, sunflowers and milk. Wheat,maize and some other products were protected against 'sudden and unexpected shocks which could harm not only the producers of these items, but also the production, market and general stability.' The Green Plan introduced a system of duties and quotas applicable to imports of agricultural products and made imports more responsive to market trends.[2]

During the 1980s, Yugoslavia passed through a time of political, social, and economic transition that changed many of its basic institutions. The world recession at the end of the 1970s caused net interest payments on foreign debts to increase considerably, net receipts from Yugoslavs working abroad to decline, and foreign lenders to withdraw. Foreign

[1] http://www.countriesquest.com/europe/yugoslavia/economy.htm.
[2] http://www.osaarchivum.org/files/holdings/300/8/3/text/80-4-374.shtml.

borrowing grew rapidly. By 1985 the deepening crisis had reduced living standards to 1965 levels. The economic crisis expanded into social, political, and constitutional crises. From 1989 through September 1990, more than a thousand companies went bankrupt. The Yugoslav wars, and consequently the loss of markets, as well as mismanagement and/or non-transparent privatization, brought further economic trouble in the 1990s.[1]

While the other Eastern European countries made progress in transition reforms and towards liberalization, Serbia was stuck in wartime conditions during the 1990s. Besides the wars in Slovenia, Croatia, and Bosnia and Herzegovina, the Serbian economy was affected by UN sanctions and the breakdown of the common Yugoslav market. The whole economic and social system acquired features of a black market which was often state-supported. Sanctions led many people to be involved in illegal business operations by hiding income, misappropriating public resources, smuggling or using smuggled goods. The shadow economy was particularly strongly reflected in the marketing of foodstuff and fuel, but also for many other goods. Hyperinflation in 1993 reached into billions of percentage points (Lyon, 1996). The large influx of refugees from former Yugoslav republics and internally displaced persons from Kosovo and Metohija (UNSCR1244) further aggravated the state economy. After a decade of civil war, economic sanctions and the NATO air war which destroyed the economy and infrastructure, Serbia's economy was only half the size it was in 1990.

The period since 2000 is characterized by rapid reforms of institutions and legislation. Economic cooperation with the EU and CEFTA countries commenced, and Serbia's application for WTO membership was another indication of the country's aim to further integrate economically with the rest of the world. Agriculture is considered to be of strategic importance to the country's further development.

Table 3.1. Historical overview of institutional development.

Period/ Motivation	Economic/ Trade/Agricultural Policies	Institutional Environment
1830/1880 Independence and creating a sovereign state	Land property rights, Agriculture, primary sector	First modern constitution: abolition of feudalism, enclosure of village forests forbidden and proclamation of the right to own land (inviolable right to private property). Multiparty system exists; industry undeveloped, with the economy relying on agriculture and mining. Extensive export of pigs and other livestock, wool, cheese and plums. First cooperative (third in the world) established in Vojvodina 1846. Schools were funded and there was a special improvement in agriculture.

[1] http://en.wikipedia.org/wiki/Economy_of_the_Socialist_Federal_Republic_of_Yugoslavia.

Table 3.1. Historical overview of institutional development.

Period/ Motivation	Economic/ Trade/Agri-cultural Policies	Institutional Environment
1880-1914 Military instability	Beginning of industrial development Political and military instability	Declaration of the Kingdom. Industry growth, emergence of labour movement and first association of agricultural producers. Expansion of the railway, huge pressure from Austria and the blockage of export (enormous taxes). Balkan Wars, strong opposition of other countries to Serbia's independence. Multiparty political system existed, but political freedom varied.
1918-1941 The development of a modern state	First agrarian reform Entry of foreign capital	In WW I, 60% of the whole male population was killed, which delayed the renewal of the country. Kingdom of Serbs, Croatians and Slovenians, later known as Kingdom of Yugoslavia. Integration of Vojvodina and Serbia. First agrarian reform and colonization of people from underdeveloped parts of the Kingdom to the north of the country. War veterans were given land belonging to the state, landlords and the royal family. Modernized industry, but mainly in northern and western parts of the country. Big influx of foreign capital (about 45%). Protection from foreign competition by introducing high custom duties.
1945-53 A centrally planned economy	Isolated country, including economy Nationalization Agriculture supported industry Food shortage	Blockage of the country by the Cominform. Intensive industrial growth and renewal of the country; radical changes in ownership structure. 1,5 million ha of land distributed to the poor. Monetary reform. Planned administrative economies. Impact of the market is completely minimized. Prescribed mandatory quotas for food products which had to be produced in households. The State was the mediator in commodity-currency exchange. Fixed salaries.
1950-1964 From central planning to a socialist self-management, collaborating with West	Partially open economy social justice and equality	State property became socially owned. Beginning of labour self-governance. The State has a mere controlling function, but most of the business decisions were made by enterprises. Membership in GATT, cooperation with OECD. Big investments in infrastructure and industry. Rate of economic growth among the highest in the world. Intense exodus of the people from rural areas. Social justice and equal opportunities. Single-party system, industrial labourers are being favoured. Growing foreign debt, cooperation with African and Asian countries (Non-Aligned Movement). Yugoslavia was the first country in the world that signed a trade agreement with the EU. Capacity building of administration and science (freedom of movement outside the borders of the country).

Table 3.1. Historical overview of institutional development.

Period/ Motivation	**Economic/ Trade/Agri-cultural Policies**	**Institutional Environment**
1965-1974 The pressures for greater political and economic freedom	Decentralization Huge support of private sector in agriculture	Strengthening of economic freedoms, evolution of the first small private businesses. Increase in social differences and unemployment. Increase of pressure for the introduction of multiparty system. Strengthening of nationalism, student demonstrations; strengthening of technocracy and bureaucracy. 1974 Constitution gave great freedom to the federal republics and the provinces; decentralization of the government and funds; intense growth of agriculture and rise of the modern food industry. Increase of political freedom, but single-party political system still present.
1974-90 Strengthening private sector and first crisis	Industrial and political freedom Outset of the first crisis	Growth of private business (SME), economic freedom, strengthening of export; tremendous support of agriculture, modernization of farm households. Land maximum still exists but the land rental is allowed. High indebtedness, monetary fluxes (inflation, devaluation). Emergence of black economy. Economic and staff related demise of institutions at the end of this period. Significant technological advances.
1990-2000 Civil War and ending the myth of the ideal country	War economy Agriculture as a means of keeping social peace	Established multiparty system. Rapid growth of nationalism, disintegration of the state, civil war, disintegration of the unique market, UN sanctions, large number of refugees, deep economic and social crisis, recession, hyperinflation (highest ever recorded in the world), significant slowdown of reforms, drastic fall of living standards, black economy, beginning of privatization and restitution; centralization of funds! Agriculture bears the burden of crisis - social peace is being maintained with low food prices. Abolished land maximum. NATO aggression, fall of the regime.
2001-present Rapid transition reforms, Integrating with international markets	Reform of the institutions and legislative Gradual liberalization of agricultural market	Prompt adoption of newly reformed laws. Economic cooperation with EU and CEFTA countries, harmonization of quality standards, privatization, gradual liberalization of local market, political instability, unresolved issues with neighbouring countries (property of exiled people, status of Kosovo, dual citizenship), fight against organized crime. Agriculture is a strategically important sector for national economy.

Source: Author's elaboration is based on: http://www.mongabay.com/history/yugoslavia, http://en.wikipedia.org, http://www.osaarchivum.or, http://www.countriesquest.com.

3.3 Natural resource institutions

3.3.1 Land institutions and implementing organizations

The first Serbian Constitution adopted in 1835 provided the opportunity of private ownership over the land and free disposal of it. This Constitution effectively abolished feudalism. In order to prevent hunger and poverty a minimum farm size was prescribed, below which land could not be seized, not even to retrieve debt. Feudalism in parts of Yugoslavia that remained under the Turkish rule ended with the agrarian reform in 1919.

By the Law on Agrarian Reform and Colonization (1945) fixed assets and other property were confiscated from their owners and became state property. 53 per cent of the land passed into the state sector or was given to peasant cooperatives; less than half, i.e. 47 per cent was given to poor farmers and colonists (new settlers that came from the underdeveloped regions). Nowadays, the Republic of Serbia has a complex system of laws, rules and procedures with regard to ownership, leasing and development of land. For the most part, property rights of land and buildings are based on the Civil Law and procedures are in place for determining land rights.

Land management in Serbia faces two key problems of institutional organization: one is the absence of adequate cadastral maps and records as well as inadequately defined land titles, and another is the slow process of land redistribution, which essentially leads to inefficient use of land resources.

The introduction of *cadastre and land registry* legislation started in all of Yugoslavia in 1928 by introducing the cadastre. The cadastre records physical characteristics of land - position, shape, size, culture, quality, class - and cadastral income users. Before 1928 a cadastre existed only on the territory which belonged to the Austro-Hungarian Empire. The law on land registry was introduced in 1931 and before that existed only on the territory of Vojvodina. Today, land registries exist in most parts of Vojvodina, Belgrade and some other big cities. The book title in its current form exists since 1930 only in municipalities that were under Turkish rule until 1912 (South Serbia). A law on real estate cadastre has been introduced in 1983. It represents unified, physical and legal records of real estate. It is used generally in small municipalities in central Serbia and Vojvodina. Owners may decide whether they want to be registered in the land register as a land owner or not. This has led to large discrepancies in the data entered in the cadastre and the land register. During the 1980s, the republics of Yugoslavia adopted rules that replaced the pre-war legislation, but the collapse of the country and the economic crisis have significantly limited the effects of this reform.

The law on cadastre and real estate rights from 1983 was amended eight times - the last time in 2005 - and remained in force until the adoption of the state survey and cadastre in late 2009. Even today there are problems with the identification of real estate and the harmonization of these records. In addition, cadastral records are not updated and the land cadastre does not contain data on current rights to real estate; it is not an ownership registry. Serbia is in the process of shifting from an analogue to a digital cadastral register (sponsored by the World Bank and other donors, see Box 3.1). The digital real estate cadastre should solve numerous problems (archiving, maintenance, distribution of the data), and contribute to a more reliable, efficient and comprehensive database on land use.

Box 3.1. Real Estate Cadastre and Registration Project supported by World Bank.

The main aim of this project is to support the development and general advancement of the real estate market in Serbia through the formation of a real estate cadastre. The Project Implementation Plan (PIP) assist the Republican Geodetic Authority RGZ as the Government institution in charge of preparation and implementation of the project. The project has two components:

Technical and operational development and support: The objective of this component is the development of a methodology and formation of a real estate cadastre, as well as its maintenance combined with the quality services extended to the customers. This component implies support to the RGZ in completion of works on the real estate cadastre in shorter delays than those envisaged, support for improvement of the system of real estate registration and delivery of quality services to the clients, as well as creation of conditions for introduction of a modern digital graphic databases.

Institutional development and capacity building: The objective of this component is to ensure an institutional development and capacity building within the RGZ, among the professionals from the private sector (mostly surveyors, geodetic specialists and similar), as well as within the institutions that are introducing data, and to which a large chunk of technical tasks of importance for the real estate cadastre will be transferred. It includes: project management and programs to develop annual business plans leading to financial self-sustainability, training of staff; public awareness and development of the full land administration legal framework.

Source: http://www.rgz.gov.rs.

The first written record about the *land consolidation* in the territory of Serbia relates to the end of the eighteenth century and is about the territory of Vojvodina. Since 1956, land consolidation measures have been applied in an attempt to stop further fragmentation of holdings and parcels and to create the basic condition for more intensive land cultivation. The main driver was the desire to consolidate landholdings in the social sector, into so called social estates. In Vojvodina, the land consolidation started in 1956, in central Serbia in 1964, and in Kosovo in the 1980s.

Up to now, the land consolidation has been carried out on 1.879 million hectares of agricultural land, or about a third of the agricultural area in Serbia. Land consolidation covers most of the countryside in Vojvodina (60 per cent of all territory in Vojvodina), followed by Central Serbia (about 9 per cent) and least in Kosovo (5 per cent). In the whole territory of Serbia, a single record of fixed assets was established on an area of about 87 per cent of the territory. In the remaining part of Serbia the ownership right is kept in the cadastral registers, books of title deeds or encumbrance books. Up to now there is no special record of agricultural parcels, but only records of cadastral parcels. The Republic Geodetic Authority (RGZ) is the organization that provides technical services and public administration related to state's survey, cadastre, basic geodetic works, the address register, topographic and cartographic activities, valuation of real estate, geodetic and cadastral information system and the National Geospatial Data Infrastructure.

3.3.2 Water resources institutions and implementing organizations

The first Water Law in force on the territory of Vojvodina was the Austro Hungarian Water Act of 1869. On the territory of the Kingdom of Yugoslavia, the Ministry of Agriculture and Water became responsible for water management. Since World War II the state law on water has been revised several times.

The main water management institution in Serbia is the Directorate for Water of the Ministry of Agriculture, Forestry and Water Management. The Directorate initiates policy, drafts laws, signs permits prepared by its implementation bodies and ensures compliance. Some competences in the field of water management were passed to the Government of the Vojvodina Province. Implementation bodies are two public water management companies (PWMC): PWMC 'Srbijavode' and 'Vode Vojvodine'. PWMCs are primarily responsible for the maintenance of water control measures, including flood protection, drainage and irrigation.

The Republic of Serbia Water Resources Development Master Plan was adopted in 2002 and is a basic document which serves to establish the basic strategy of water utilization, water protection and protection against flooding in the territory of the Republic of Serbia.

3.3.3 Forest institutions and implementing organizations

A Forestry Development Strategy of the Republic of Serbia was adopted in 2006. In this strategy document the government has designed basic principles, development goals, measures and ways to support the sector's development. With this strategy, the government has set a clear framework and requirements for institutional reforms, redefining roles and responsibilities in the entire forestry sector (Nonić and Milijic, 2008).

The Forestry Directorate is an administrative body within the Ministry that performs the tasks of state administration and activities related to forestry policy, forest preservation, promotion and utilization of forests and wildlife, and enforcement of measures to protect forests and wildlife. A public enterprise 'Serbiasume' is responsible for the protection, promotion, exploitation and management of forests, forest land and other resources of forests. Forest management in the territory of Vojvodina is the responsibility of the public company 'Vojvodinašume'.

Private forest owmers are organized in associations to represent their members' economic interests or in a forest community, with the aim of jointly managing the forest.

3.3.4 Environmental protection institutions and implementing organizations

Environmental protection was among the basic principles of the Constitution of 1974. The increased concern and activity of the international community in matters of the environment had a direct influence on environmental legislation in Yugoslavia. The dissolution of Yugoslavia put an end to this phase of environmental protection. Serbia adopted a law on environmental protection in 1991.

Today in Serbia the bodies involved in environmental issues range from the municipality level to the state level. Based on the Law on Ministries, duties in the field of environment, on the national level, fall under the jurisdiction of various institutions. The Ministry of Environment and Spatial Planning (MESP) is the most important, as it performs duties related to environmental protection and the protection of the ozone layer, monitoring climate changes, identification and protection of natural systems that are of special importance, etc. Certain jurisdictions in the field of environmental protection were transferred to the autonomous province Vojvodina in 2002.

In local governments, municipalities and towns, corresponding departments and secretariats are in charge of dealing with environmental protection. On a local level, secretariats for environmental protection have limited responsibility in environmental issues, which includes air quality protection, municipal waste management, urban planning, construction permits for small buildings, the strategic assessment of plans and programs, the assessment of project impact on the environment and issuing permits within their competence.

For some aspects of the environmental legislation such as water management, responsibilities reside with other ministries than the MESP. Notwithstanding, the latter retains the ultimate responsibility for the negotiations with the EU for the entire Environmental Chapter. Within MESP, the Serbian Environmental Protection Agency (SEPA) is responsible for the national environmental information system, for the collection of environmental data and reports on the status of the environment. SEPA is responsible for cooperation with the European Environmental Agency (EEA) and European Information and Observation Network (EIONET).

3.4 Knowledge and innovation system

3.4.1 Education system

The Serbian national education system is characterized by the traditional structure of a voluntary pre-school education, compulsory and uniform elementary education (8 years), and an optional four-year secondary education. The higher education system has two types of study: academic study at universities, and vocational profession-oriented study at vocational colleges. Postgraduate studies are carried out through masters and doctoral programs; their requirements differ, depending on the faculty. Serbia joined the Bologna process in 2003 and subsequently initiated a gradual process of reform which received its legal support in 2005 by adoption of a new Law on Higher Education. This law formally introduced the European Credit Transfer System. From 2007/8 all new students study under the reformed study programs at all higher education institutions.

The private sector in the education system exists since early 1990. Private primary and secondary schools are permitted by law since 2003. Until now there are just a few private schools, and only in big cities. Today, in addition to seven universities founded by the Republic of Serbia, with about 182.000 students, there are ten private universities with about 35.000 students.

Serbia spent 3.5 per cent of GDP on education which is significantly below the desirable 6 per cent recommended by UNESCO and the average of OECD countries. Countries in the region allocated about 4 per cent of GDP for education. The percentage of young people attending secondary and higher education is permanently increasing. However, 20 per cent of children do not follow secondary education, especially boys and young people from marginalized areas (LSMS, 2008). The imbalance in gender at higher education institutions has increased during 2000s by a growing number of female students. Yet, the right to quality education is still not available to all children. Apparently education does not help to reduce gaps between children from different social and economic backgrounds; these gaps appear to widen during the course of education, instead of being narrowed.

3.4.2 Science and innovation system

Science and innovation policies have had significant national attention since the time that Yugoslavia existed as a state. Institutionally, responsibility in this area has changed and merged with other departments of the government (culture, education, environment, etc.), which is the case today. Increased investment in science began during the 1960s, when the figures showed a technical and technological backwardness compared to developed European countries. In the 1990s, however, there were no investments in the maintenance of equipment and its updating, and the financing system was not transparent; but the biggest effect of the crisis was felt in the extremely high rates of migration of most educated staff.

In Serbia the Ministry of Science and Technological Development (MSTD) is responsible for managing, planning and financing public R&D activities. MSTD also has core and full responsibility for international R&D cooperation.

On 25 February 2010, the Government of the Republic of Serbia adopted the Strategy of Scientific and Technological Development for the period 2010-2015. Seven priority areas are identified in the national science strategy: biomedicine, environmental protection and climate change, energy, agriculture, food science and ICTs. Most of the planned new funding will go to these focus areas. The *Serbia 2020* road map identifies low levels of investment in research, education, and technology as 'a major problem' of the country. In terms of science and education funding, Serbia is lagging behind other European countries, including regional neighbours, which, apart from Albania, all invest more in science, many over 1 per cent of GDP.

Table 3.2. EU and Serbia: the current state and development goals in their respective Science and Technology 2020 road map.

	EU 2010	EU 2020	Serbia 2010	Serbia 2020
Employment of the population 20-64 (%)	68	75	49	65
Investment in Research and Development (% of GDP)	1.9	3.0	0.3	2.0
Share of renewable energy in total energy consumption (%)	16	20	12	18
Energy efficiency (toe/$1000 GDP)	0.21	0.17	0.96	0.57
Population aged 30-34 with a university degree (%)	31	40	21	30
Poverty rates (*below 60%* of the national *median* equivalent *disposable income*)	16	12	17	14

Source: Serbia (2020), SORS.

Serbia 2020 further calls for creating better public-private research links through various tax and other incentives, which could provide scientists and engineers with career opportunities in private companies as well; this is something that is currently chronically lacking in Serbia. According to the UNESCO (2010), only 1.4 per cent of Serbian researchers work in the private sector.

3.4.3 Transfer of knowledge and innovation in the agri-food sector

Research policy, priorities and implementing agencies

New knowledge and technologies in agriculture and in the food sector is for the most part generated by scientific research funded by the MSTD of the Republic of Serbia and the Secretariat for Science and Technological Development of AP Vojvodina. The research topics and methods of work are usually determined by personal affinities of project leaders. The real needs and priorities of the agro-food sector and the impact of technological development is too rarely a decisive factor in project formulation or the formation of research teams.

In the MSTD's Co-financing Integral and Interdisciplinary Research Programme (IIRP), researchers are invited to run projects within the seven national priorities in the science domain as a response to topics laid down in the Research Cycle Programme 2011-2014. Scientific priorities are laid down in the strategy on the basis of the criteria such as: possibility of successful participation in the Lisbon strategy and scientific priorities of the European Union; budget investment amount in the last seven years; success of previous research; possibility of application in economy, in the country and abroad; contribution to successful conduct of public policies on the domestic and international levels; etc. The theme on agriculture and food sector is one of seven national priorities in IIRP, with tree defined topics, which are biotechnology in breeding, biotechnological approaches and alternative measures for the purpose of promoting competitiveness, quality and food safety, and technologies for producing new and traditional products.

Since 2009 the Ministry of Economy and Regional Development implements an innovative program for SMEs supporting them with direct financial support for the implementation of innovative activities. It also views the development of cultural SMEs investing in innovation. The main objective of the program, however, is to connect small and medium enterprises with scientific research organizations, universities, consultancies, and other centres of knowledge and technology.

National agricultural education and research system

The national agricultural research system comprises three groups of research and development organizations: public financed research organizations, state financed and private universities.

The list of registered (accredited) public financed research-developmental institutes within the MSTD includes nine institutes dealing with specific topics of agriculture: for animal husbandry, vegetables, plant protection, food technology, etc. Most of these institutes have modern equipment and good teachers. On the other hand, the deficit of personnel (due to the large emigration of young educated people) may soon become a serious problem. Besides these research and development institutes, the Ministry of Science registered only a few other institutions (about ten). These institutions do not meet accreditation standards and do not have the status of research institutes, but they are still active.

The there are universities governed by the Council of Higher Education. Serbia has a comprehensive agricultural education system, which is organized through a number of agricultural vocational schools and faculties of agriculture in the universities. The whole system is under the authority of the Ministry for Education, which is responsible for the design and the implementation of the curricula. Nowadays there is a network of 34 vocational agricultural schools evenly distributed throughout the country. Apart from the Agricultural Faculty in Belgrade, there are several other faculties of agriculture (in Novi Sad, Pristina, and Cacak), and 3 colleges (Šabac, Prokuplje and Negotin) in Serbia. There are Faculties of Veterinary Medicine in Belgrade and Novi Sad as well. Most of these faculties have various units or departments specialized in agricultural sciences (crop science, zoo technique, food processing, horticulture, land management, agro-economics, etc.).

In addition to the state agricultural faculties, there are also private universities that educate professionals of different profiles related to agriculture (Faculty for Biofarming-Sombor, Advanced Agricultural School in Backa Topola and a great number of their departments in various Serbian cities). These departments have developed special curricula meeting the needs of agricultural experts through their specific education and training requirements. These institutions do not have facilities for scientific research work in fundamental areas of agricultural sciences.

National extension system

The development of agricultural extension service in Serbia started at the middle of the 19th century. The first institutions were: National Economy (founded 1851 in Topcider) and Stud Farm (Ljubicevo 1852). In the following years there began to be organized exhibitions of foreign breeds of livestock, the establishment of state nurseries for the breeding of fruit trees and vines, the establishment of district agricultural stations, etc. A special contribution to the development of agricultural services in Serbia was the founding of the Farmers' Society (1869-1946), later transformed into the Serbian Agriculture Association. This association significantly contributed to the establishment and popularization of agricultural cooperatives and specialized associations of farmers. During the 1950's the process of revitalization of agricultural extension services began in order to perform the 'transfer between science and practice.'

Since 2010, in compliance with the Law on Advisory and Extension Services in Agriculture,[1] the Agricultural Extension Service has been operating on two levels: at the Republic level, under the responsibility of MAFWM, for those regional centres located outside the territory of the Autonomous Province of Vojvodina, and at the provincial level, in the territory of the Autonomous Province of Vojvodina, under the responsibility of the Provincial Secretariat of Agriculture, Water Management and Forestry.

The Law on Agriculture Extension Service (2010) aims at improving the regulated advisory system that informs farmers about modern standards of product quality, farm management and environmental protection. The law defines the advisory expertise and, in detail, the extension activities that those experts can provide to the Serbian farmers. It also regulates the issuing of licences to the extension experts (through the Ministry of Agriculture) and defines the training that extension experts must complete as well as the system of funding through the Serbian and Vojvodina budgets. The agricultural extension service network includes 34 regional centres (22 in Central Serbia and 12 in Vojvodina), employing about 230 extension agents. A special form of extension services is to monitor production in selected farms where extension agents record the costs of inputs, yields and other data related to farm management. Extension agents, in addition to their work conducted on farms, are available to other households, giving them advice and recommendations, information about new technologies and new methods of farm management. Advice, assistance and other activities that producers receive from extension agents are free of charge to all beneficiaries.

Individuals, institutes and private domestic and foreign companies also provide consultancy service in the agro food sector through rural networks and consulting companies. A system of NGOs that is supported by the MAFWM disseminates knowledge and information in rural areas under the name of the Rural Network. The network consists of regional centres for rural development and of local action groups established at the

---

[1] Official Gazette No. 30/10.

municipal level. The network began operations in 2004 and has been actively involved in training programmes for local rural development actors. Input suppliers provide corresponding extension services in order to promote their products, thus ensuring the products' proper use and to maintain the firms' market share. So do pharmaceutical companies, *equipment manufacturers* and commodity dealers and many others. Such consulting services are often directed by commercial interests, exposing the farmers to the risk of contradictory information and advice.

3.5 Financial institutions

3.5.1 Credit market

Various models were used in former Yugoslavia to finance agriculture and rural/regional development. However, there has never been a stable model of agricultural financing, even before transition. Since in the Communist epoch the Government had control of the banks, it had the leading role in defining both short-term and long-term models of agricultural financing. Scarce financial resources were allocated often on the basis of non-economic principles and without transparent procedures. Socially owned enterprises and cooperatives had privileged positions with regard to access to credits (see section 3.2). Agriculture has lost the status of a privileged sector by reconstruction of the financial system (in 1994). Since that time the financing of agriculture through loans from the National Bank was abandoned and agriculture had to rely on commercial banks for loans.

In 2004, in order to enhance competitiveness, modernize farm machinery and equipment after more than a decade of crisis, the MAFWM established a model of short- and long-term loans for agriculture, under more favourable conditions than the market ones. Long-term loans were granted to registered agricultural holdings by commercial banks for specific purposes: building and purchasing of irrigation systems and equipment, purchasing of agricultural machinery, establishing plantations of perennial plants, establishing glass houses and greenhouses, as well as investing in livestock production. The so called agricultural budget credit funds were distributed through the Development Fund of the Republic of Serbia, partly through commercial banks for short term and long-term loans, and partly directly through the Development Fund to agricultural processing facilities. Use was limited to registered agricultural holdings (natural and legal persons).

In 2009, a new model of long-term loans was established in which banks participated for 60 per cent of the amount, and the Ministry for 40 per cent. The bank's fund is to be repaid in the first three years of the repayment period. Interest is calculated according to the bank's business policy. The 40 per cent for the amount of the Ministry are to be repaid in the two subsequent years and no interest is charged. So in practice the interest on the total amount of a long-term loan is 5 per cent per annum. The same model of long-term loans applies also in 2010 and continues in 2011.

Table 3.3. Credits disbursed from the agrarian budget (1,000 euro).

	MAFWM		Provincial Secretariat for Agriculture, Forestry and Water Management	Total amount
	Short-term loans	Long-term loans		
2004	29,916	21,066	0	50,981
2005	20,504	24,123	0	44,627
2006	39,379	10,717	238	50,334
2007	18,729	4,994	147	23,870
2008	20,665	17,841	1,410	39,916
2009	22,490	24,437	0	0

Source: GoRS (2011).

In 2010, the Ministry also channelled credit support through local government and the Development Fund of Vojvodina. These models of lending in cooperation with local governments are also executed through commercial banks, with the Ministry and local governments contributing with 25 per cent of the amount, on which no interest is calculated, and 50 per cent of bank resources on which the bank calculates interest according to its business policy. Loans to the registered agricultural households are approved by the bank in cooperation with the local government.

3.5.2 Development funds

Several funds have been mentioned above that help providing credit to the agri-food supply chain. The Development Fund of the Republic of Serbia was established in 1992. The resources are used for financing programmes in the field of economic and regional development, the encouragement of small and medium-scale enterprises, financing export opportunities, financing the economy for liquidity and financing regular Fund activities. The credits offered from the resources of the Fund can either be short-term or long-term credits, where short-term credits were often used for encouraging exports and long-term credits for financing investments by small and medium-scale enterprises with the aim of encouraging the harmonization of regional development.

In 2005, the Fund for stimulation of the development of agricultural production in Serbia was established as an account in the Treasury general ledger. The Fund was established to encourage the development of agricultural production by providing short- and long-term loans to registered agricultural holdings via commercial banks. The Fund is financed from the following sources: funds provided in the budget of the Republic of Serbia for the current year, funds generated from the repayment of loans and other means. In addition to the republican Fund, there is also a fund for agricultural development established by the AP Vojvodina.

3.5.3 Leasing companies

The development of the system of financial leasing started with the adoption of the Act of the Republic of Serbia regarding financial leasing in May 2003 ('Official Gazette of the RS', Nos. 55/03 and 61/05). The establishment of the domestic leasing enterprise 'Lipaks' was initiated in November 2003 by the Government of Serbia, with the objective of ensuring favourable conditions for the purchase of tractors, agricultural machinery, trucks and specialized vehicles of domestic producers. The major users of leasing facilities, or agricultural machinery, were private farmers (because they were offered the restoration of dilapidated agricultural machinery under relatively favourable conditions). In June 2004, nine leading leasing companies in Serbia founded the Association of Leasing Companies (ALCS). The foundation of the Association had been initiated by the SEED Programme of the World Bank for the enhancement of small- and medium-scale enterprises in Southeast Europe. Since 2005 the conditions for leasing agricultural machinery have been favourable with a somewhat lower interest rate than for other leasing facilities.

Evaluation: use of financial services in the agricultural sector

Since 2005 financial services in Serbia have grown and improved significantly. At the beginning there were the standard constraints such as limited trust in the banking sector and other financial institutions among the farmers (as a result of previous bad experience from 'pyramidal' schemes forms in the 1990s), lack of experience and expertise in writing and presenting business plans (also the extension service had insufficient expertise to help farmers), too high interest rates (usually indexed to the euro), collateral problems (unfinished cadastre, non-existing credit history), etc. Many of these problems have been overcome in the meantime, also due to higher engagement of banks and advisory services on the farmers' education. The biggest limitation is, however, market volatility, which makes farmers unsure and reluctant to large investments.

Sources for financing the Serbian agricultural sector are the agrarian budget, business banks, leasing companies and development funds (state, provincial and municipal). However, although they have enormous significance for the overall development of agriculture and rural areas, especially in the transition period, these sources and/or the credit conditions were not able to match the sector's needs for credits (Bozic *et al.*, 2009).

Looking at some figures on loans provided to the agricultural sector indicates that credits for farm mechanization are by far the most important – this item shows the highest number of approved applications and the greatest share in the total amount of credit (see Figure 3.1). Regional differences in the number of approved loans in all observed years indicate the dominant position of Vojvodina. This situation can be explained by better information, better institutional preparedness (land registry), and a larger number of market-oriented producers in Vojvodina. The average amount of granted loans indicates that these are significantly higher in Vojvodina. The reason is not only the larger farm size in Vojvodina, but also the fact that the biggest agricultural enterprises, which are also entitled to subsidized loans, are located in this region.

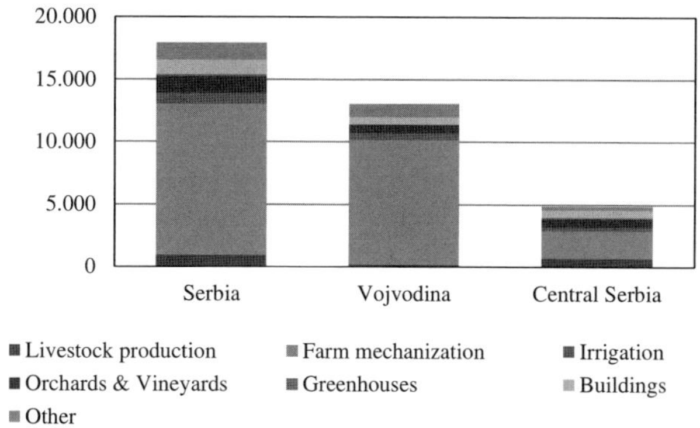

Figure 3.1. Released long-term loans in 2008 (1,000 euro).

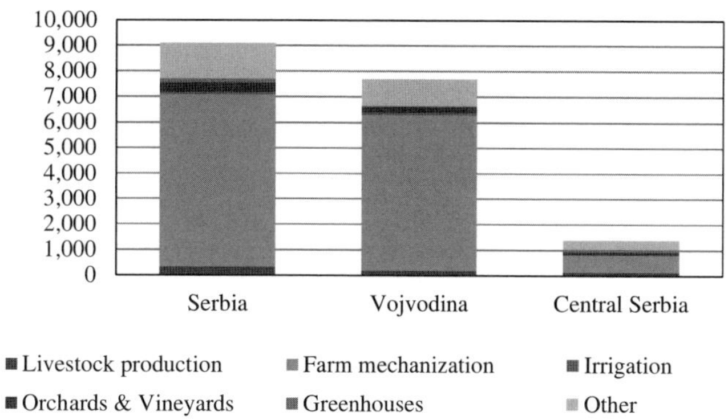

Figure 3.2. Released long-term loans in 2009 (1,000 euro).

3.6 Cooperatives and farmers' organizations

3.6.1 Cooperatives

The cooperative concept in Serbia dates from 1850's, with a first cooperative established in Vojvodina in 1846. The development of cooperatives has enabled the owners of small farms to expand their business, including land purchase on favourable credit terms. Up to 1900 the agricultural sector in Serbia contained over 600 cooperatives. After WW-II and the brief experiment with central planning and collectivization of agriculture according to the Soviet model, the Yugoslav government adopted in 1950 a radically new economic model. The system of worker self-management introduced important elements of economic democracy and participation in the industrial structure. In the agricultural sector the old agricultural cooperatives were modified and partially incorporated into the system of worker self-management. The key change was the effective abolition of collective property -- the law recognized only the 'social' and private property. Finally, unlike the situation for cooperative enterprises in Western Europe, the Communist Party retained the right to interfere in the activities of the supposedly independent self-governing enterprises.

The situation in the cooperative sector in Serbia today could be described as very unfavourable. The process of post-socialist transformation had very negative effects. The status of collective property remained too long unregulated, which significantly reduced the chances to effectively adapt to new market conditions. Reliable information on the scope of business of the cooperative sector and its functional units is lacking. Many fields of agriculture which previously had been dominated by cooperatives (dairy sector), and thus secured the fair income of primary production, collapsed.

Official data on the number of registered cooperatives vary according to the source: according to the SORS, there are about 820 registered companies and cooperatives. According to the Serbian Business Registers, there were more than 1,600 active agricultural cooperatives in early 2011. Many of these cooperatives are marginal: organized as cooperatives and in reality counting no more than a few members. The oldest agricultural cooperatives in Serbia are also the largest. These cooperatives were established to support rural communities in order to reduce poverty and underdevelopment in remote areas. Cooperatives that were established after 2000 are often associated with foreign donor projects so that they are mostly (still) small and with insufficient capital. Many of these cooperatives were the fruit of donations or are encouraged by other means.

It is estimated that the agricultural cooperatives in Serbia are in relation with about 115,000 subcontractors, while the number of cooperative members and employees is less than 30,000. Sub-contractors contribute to the successful business of the cooperative, but they do not to take part in sharing the income and benefits provided by cooperatives. Often a small group of employees and members of the cooperative has taken over the property of a given cooperative, wanting to maximize profit. On the other hand, they deny that possibility to the farmers who regularly cooperate with the cooperative and who should actually be given the chance to become members.

The Cooperative Union of Serbia (CUS) is the voluntary association of cooperatives. It operates as an independent non-profit organization. It provides, *inter alia*, assistance to the members concerning the legal status of cooperatives, reviews the market situation, organizes competitions (agricultural record-breakers), and cooperates with other institutions and state bodies in matters of agricultural policy. The Union helps in developing business plans, in making connections with financial institutions, and in other issues of interest to the work and operation of agricultural cooperatives. It has 14 regional units and does not have funds of its own or access to the financial market. The privatization process left CUS without significant property of its own.

Although cooperatives have a long tradition in Serbia, their development has been hampered by legislative challenges, insufficient support mechanisms and other factors. Serbia currently has a Law on Cooperatives from 1996, which was last amended in 2006, and a draft Law on Cooperatives prepared during 2010 but not adopted yet. The draft law defines cooperatives as 'legal entities that stand for a special form of organization of cooperative members who, by carrying out business activities on cooperative principles, realize their economic, social and cultural interests, and control the cooperative in a democratic manner'. The new law will create an appropriate regulatory and legal environment for reforms in the co-operative sector in order to create the necessary conditions for it to become an important and dynamic factor in the development of the economy.

3.6.2 Association for agriculture, food and tobacco industry and water management of the Chamber of Commerce

This association, a special organizational unit within the Chamber of Commerce aims to follow and analyse the trends in the fields of agriculture and food industry, propose improvements in the work of members of the associations, initiate economic and agricultural policy measures and address the competent institutions, to take part in deliberations on the protection of domestic production, propose measures to combat monopolies on the domestic market and to organize international fairs and promote local agriculture at the fairs.

Within this association, groups have been set up of producers of primary agri-food and fishing products (19 in total). The meetings within these groups are attended by the representatives of producers of primary agricultural products, of the manufacturing industry and by representatives of the Ministry. On that occasion they present the current situation in a particular area of agriculture, suggest possible solutions, and initiate proposals for agricultural policy measures for the next year and so on.

3.6.3 Business associations and farmers' associations

A number of associations at the national, regional and local levels are involved in the development of agriculture and rural development. As part of their activities, these associations organize conferences and thematic lectures where the participants are experts and extension agents from the country for the specific area or from abroad. Networking of producers' associations of different profiles of activity has started, but a central institution at the national level does not yet exist. Many of such associations keep a low profile and are only occasionally present in public decision making processes. Producers' associations are often associated with political parties, receiving their tacit or transparent support.

3.7 Agricultural policy implementation institutions

3.7.1 The MAFWM

The Ministry of Agriculture, Forestry and Water Management (MAFWM) is responsible for proposing system solutions in the areas of agriculture, for the definition of agricultural policy and for its implementation. The Ministry proposes to the Government the program of agricultural budget allocations under the national budget, as well as other acts with respect to the functioning of the agricultural sector (see also chapter 7). It also proposes legislation, monitors and analyses agricultural policy measures and supervises law enforcement, and is responsible for inspection services.

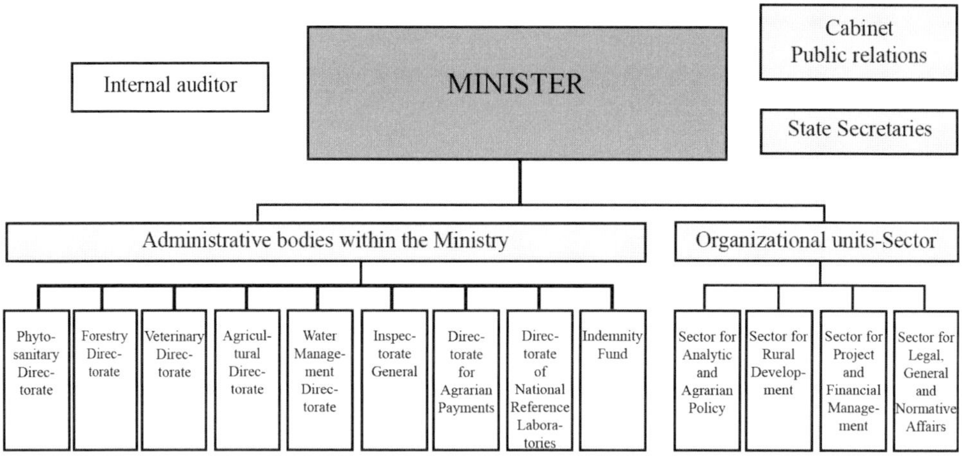

Figure 3.3. Organogram of the Ministry of Agriculture, Forestry and Water Management of the Republic of Serbia (2010).

3.7.2 Adapting agricultural institutions to the EU model

To prepare for the use of EU's Instrument for Pre-Accession Assistance (IPA, which consists of five components – see EU Enlargement website for more detailed information), the MAFWM and the Serbian government have deployed a number of activities. One is to establish the legal framework and another is to put in place the organizational and operational structure of this policy support programme.

The *Law on Agriculture and Rural Development* (Official Gazette of RS, No. 41/09) established the necessary legal framework, creating the possibility to receive assistance through IPA for component V Rural Development. This law defines a number of objectives of agricultural policy and how to achieve these, the types of incentives available for agricultural development, conditions for exercising rights to incentives, and the potential beneficiaries. This law also lays down rules on the establishment of a Register of Agricultural Holdings (see textbox 3.2), and the rules on record keeping and reporting in agriculture, as well as on an integrated agricultural information system. According to this law an Agrarian Council composed of the representatives of agricultural holdings and family farms, farmers' associations and other participants in the chain of agricultural production will be established.

General elements of the operational structure of the IPA are the National IPA Coordinator and the Competent Accrediting Officer (CAO). Parts of the control and management systems are the National Accrediting Officer (NAO) and the National Fund. The *National Authorizing Officer* (NAO) and the *Director of the National Fund* have been appointed in the Ministry of Finance. They are responsible for national accreditation of the system, receiving money from Brussels for delivery to the Paying Agency (in the case of Serbia, to the Directorate for Agrarian Payments within the MAFWM). The last and most operative parts of the system are the Paying Agency and the Governing body, responsible for the programming of assistance (IPARD preparation) and the payment to final users.

The operating structure for the implementation of the rural development component of the IPA (IPARD) programme includes the Managing Authority and the Paying Agency.

They are under the jurisdiction of the MAFWM. Tasks of the Managing Authority (MA) are performed by the Working Body for Management of the EU Rural Development Programs, which has officially been established on 21 October 2010, as a part of the Rural Development Programme.

At the present time, the task of the Paying Agency (PA) is performed by the Directorate for Agrarian Payments (DAP), established in October 2009 within the MAFWM (see Figure 3.3) and responsible for payments in agriculture. The DAP currently performs the implementation of national support measures, and after the EU accreditation for the use of funds from the rural development component of IPA fund it will be responsible for the implementation of IPARD programme.

Since 2010 there is a twinning project - 'Strengthening the capacities of Serbia to receive the resources from the EU funds intended for rural development in the pre-accession period'. The project aim is to establish the capacities in the Payment Agency and the Managing Authority. The backbone of the entire administration is an integrated software system that is the basis for the implementation of two pilot IPARD measures, and that will be the basis for implementation of all IPARD measures after the accreditation. In the forthcoming period this system will be the basis for the development of IACS (Integrated Administration Control System).

There are still some parts of the EU supporting system to agricultural policy -- such as a market intervention agency, farm accounting system and a land parcel identification system -- that have not yet been established or are only in the process of transformation. The Republic Directorate for Commodity Reserves under the MAFWM acts like a kind of market intervention agency. If there are indications of market disturbances that can to put at risk the functioning of the market and food security of the country, the Directorate intervenes through purchasing or bringing out certain amount of products on the markets (Zivadinovic and Milovanovic, 2010). Such kinds of interventions are implemented based on a Government decision. In Serbia, there is no system of production quotas, and export taxes are not applied either.

The Farm Accountancy Data Network (FADN) is a European system of sample surveys conducted every year to collect structural and accountancy data on farms, with the aim of monitoring the income and business activities of agricultural holdings and evaluating the impact of the measures taken under the Common Agricultural Policy (CAP). A FADN system does not exist in Serbia, or any other database of farm performances that can provide information in the same way as the FADN system. The Law on Agriculture and Rural Development of the Republic of Serbia (Article 33) stipulates that the MAFWM keeps a system of agricultural accountancy data to monitor the level of income and expenditure of the registered farms and family farms, in order to assess the efficiency of agricultural production and to analyse the agricultural policy measures. The establishment of FADN will be financed from the IPA 2010 programming cycle, and the implementation of this project is expected to start in 2011.

A Land Parcel Identification System (LPIS) is one of the main priorities for MAFWM. It concerns the establishment of an agricultural land information system in Serbia, which is necessary to implement agricultural support policies according to the EU *acquis*.

Box 3.2. Register of agricultural holdings.

Since 2004, the entry into the registry is operated by the MAFWM in cooperation with the Ministry of Finance of the Republic of Serbia - Treasury Administration, which has the necessary infrastructure throughout the Republic of Serbia for the daily updates. The Directorate for Agrarian Payments, having been set up as an independent unit within the Ministry, had a good case to be also the keeper of the Register of Agricultural Holdings. However, due to lack of financial resources the decision was made for the register to remain in the Treasury Administration of the Ministry of Finance, while the owner and user of information is the Directorate for Agrarian Payments.

The current Rulebook on making entries and maintaining the register of agricultural holdings (Official Gazette of RS, Nos. 111/09 and 21/10) defines an agricultural holding to mean a holding with at least 0.5 hectares of agricultural land in the territory of the Republic of Serbia where farmers, companies, cooperatives, other legal entities such as institutions, schools, monasteries, churches and other organizations and entrepreneurs are engaged in agricultural production. A holding with less than 0.5 hectares may be entered into the Register if the holder is engaged in livestock, vineyard or vegetable production (glasshouses and greenhouses) or other forms of agricultural production (fish farming, cultivation of mushrooms, snails, bees) on it.

The agricultural holder is to be entered into the Register and when a farmer, subject to registration, is a natural person, members of the family farm may also be entered. The plots are entered in the Register on the basis of written statements from the single real estate records of the Republic Geodetic Authority (cadastre and land registry).

Every year a renewal of registration is done at the request of the holder. Agricultural holding in the registry may have an active or a passive status. This is important from the aspect of the beneficiary of budget funds, as only an active holding is eligible for benefiting from the incentives.

The status of an active holding is lost, *inter alia*, in the case of non-compliance with the regulations relating to: discharges and disposal of hazardous and harmful substances on agricultural land and in drainage canals, failure to exercise control of the fertility of arable land and not keeping records of the quantity of mineral fertilizers and pesticides introduced in the arable agricultural land designated in cadastre classes one to five, and on animal identification. Respect for these rules provides the agro-environmental standards, basic for future implementation of the cross compliance measures, which is a condition for the use of financial incentives. Recuperation of the status of an active agricultural holding takes from one to three years depending on the type of regulation that was violated.

Source: GoRS, 2011.

3.8 Conclusions

This chapter describes the current situation of institutions related to the agricultural sector in Serbia. The institutional system in Serbia has been heavily affected by the frequent changes of state borders, and the socio-economic and political system. These changes have brought radical shifts in the legal, organizational and implementation mechanisms related to agricultural policy. As in other communist countries, after WW-II the institutional system was deeply influenced by the ruling political ideology. An important feature of the institutional structure is the fact that Vojvodina has developed its own institutions at the time of the Austro-Hungarian monarchy, and that in this part of the country institutions are better organized and have a longer tradition.

Substantial reforms of institutions and legislation started after the political changes in 2000. Nevertheless, according to the EBRD, the progress that Serbia has achieved in structural and institutional reforms is significantly behind other countries in transition. From the beginning of the transition a major challenge in the field of agriculture has been to combat the grey economy and to establish an efficient system of border controls. Fundamental institutional reform related to land, cooperatives, advisory services, laboratory system and so on has been slow causing a lack of substantial structural reforms in the sector. The reason for this slow reform process can be linked with a deficiency of inter-ministerial cooperation and an extensive overlapping of responsibilities between different institutions.

In view of EU membership, the Serbian government have been occupied with the modernization of administrative arrangements, whereas beforehand it made little effort to develop administrative and social capacities. The system of transfer of knowledge and new technologies is reformed, but not yet effective and matched with needs of small farms. Producer associations, associations, horizontal and vertical networking of business sector are insufficiently and unevenly developed.

Significant progress has been made in the area of institutional building and preparation of strategic documents related to the use of pre-accession EU funds. Nevertheless, the major constraint remains the deficiency of competent staff. Improving the human resources of the administration should have priority, as well as better coordination between the various governmental bodies.

Literature and websites

Bozic, D., P. Muncan and N. Bogdanov, 'Credits for Serbian agriculture.' In: *113th EAAE Seminar, The role of knowledge, innovation and human capital in multifunctional agriculture and territorial rural development.* Belgrade, 2009.

GoRS (Government of the Republic of Serbia), *Answers to the European Commission's Questionnaire.* Belgrade, 2011.

LSMS, Living Standards Measurement Survey 2008. Available at www.worldbank.sr

Lyon, J., 'Yugoslavia's hyperinflation, 1993-1994: a social history.' In: *East Auropean Politics and Societies* 10 (Spring, 1996) 2: pp. 293-327.

Nonić, D. and V. Milijic, *Analysis of the current state of the private forestry sector in Serbia and its role in the process of defining the Forestry Development Strategy and the National Forest Programme (in Serbian).* PROFOR. Belgrade, 2008.

Serbia 2020. *The concept of Serbian development by the year 2020.* Draft for public debate. 2010. Available at www.vojvodinahouse.rs

UNESCO, *UNESCO Science Report 2010. The current status of science around the world.* UNESCO Publishing, 2010

Zivadinović, B. and M. Milovanović, *Guide through EU policies – Agriculture (in Serbian),* European Movement in Serbia, Belgrade, 2010.

Web sources

www.mongabay.com/history/yugoslavia
www.en.wikipedia.org
www.osaarchivum.org
www.countriesquest.com
www.rgz.gov.rs

Chapter 4

4 Agricultural production, prices and income

4.1 Introduction

This chapter provides an overview of the agricultural sector in the Serbian economy, covering land resources, farm structure, production, prices and income. The most important indicators of the agricultural sector in Serbia are compared with other Western Balkan countries (including Romania and Bulgaria) and EU. The second section outlines the role of agriculture in the general economy, showing its shares in GDP, employment and foreign trade. Sections 3, 4 and 5 assess the land resources, farm structure and land ownership. Section 6 covers the size of agricultural production, its diversity and economic value, with details about most important sub-sectors. Section 7 looks at output price indices for agriculture, followed by trends in real prices for specific agricultural products. Section 8 concludes this chapter with a brief evaluation of the income situation in Serbian agricultural sector.

4.2 Agriculture in the economy

The contribution of agriculture to the national economy in Serbia is significant. The relatively high share of the agricultural sector in the country's GDP and employment is due to a slowly progressing restructuring of the rest of the economy, overall low investment activity and consequently low employment opportunities in non-agricultural sectors. Over the last decade the share of agriculture in GDP decreased from over 14 to about 10 per cent, whereas the share of agriculture in employment remains over 20 per cent (see Table 4.1).The importance of agriculture in Serbia is still significantly higher compared with other countries in the region, except for Albania (see Table 4.2).

Table 4.1. Indicators of the agricultural sector of Serbia.

	2001	2002	2003	2004	2005	2006	2007	2008	2009	2010
% of Agriculture, hunting, forestry, fishery in GDP	14.3	13.3	12.1	13.2	12	11.4	9.8	10.1	10.6	10.3
Employees at Agriculture, forestry and water works supply (x1000)	n.a.	n.a	n.a	642	637	540	553	707	623	533
% of Agriculture, hunting, forestry, fishery in total employment	n.a.	n.a	n.a	23.9	23.2	20.5	20.8	25.1	23.8	22.2

Source: SORS, Statistical Yearbooks, LFS.

Table 4.2. Importance of agriculture in Serbia, neighbouring countries and EU-27.

	Share in GVA (%)		**Share in employment (%)**	
	2000	**2008**	**2000**	**2008**
Albania	25.5	18.5	71.8	57.4
Bosnia Herzegovina	11.8	9.1	19.8	20.6
Croatia	8.4	6.4	16.9	13.2
FYR Macedonia	12.0	10.8	27.0	18.2
Serbia	19.7	12.0	23.9	21.4
EU-27	2.4	1.8	n.a.	5.4

Source: Eurostat, World Bank.

Agriculture contributes significantly to the country's trade balance. Since 2001, when the agri-food sector had a negative trade balance, exports have grown at a higher rate than imports, resulting in a positive trade balance since 2005 (Table 4.3). The most important agricultural export products are maize, beet sugar, frozen raspberries and wheat. Imports are dominated by fruits, vegetables, coffee, tea, cocoa and spices. Exports of Serbian agriculture have benefitted from a preferential status in trade relations with the EU (see also chapter 8) and trade liberalization with the Western Balkan countries (CEFTA). The share of agricultural exports to total exports is significant, and also agricultural imports account for a considerable share of the country's total imports (Table 4.3).

Table 4.3. Importance of agri-food trade.

	Units	**2000**	**2001**	**2002**	**2003**	**2004**	**2005**	**2006**	**2007**	**2008**	**2009**
Agri-food exports	million euro	320	347	555	509	629	732	992	1218	1339	1397
Agri-food imports	million euro	311	507	578	577	687	622	721	818	1004	940
Agri-food trade balance	million euro	9	-160	-23	-68	-58	110	271	400	335	457
Share of agri-food trade in:											
- Total exports	%	19.0	18.3	25.3	20.9	22.2	20.3	19.4	18.9	18.0	23.3
- Total imports	%	8.6	10.7	9.8	8.8	8.0	7.4	6.9	6.1	6.5	8.4

Note: Agri-food trade according to Combined Nomenclature of Custom Tariffs (CNCT).
Source: Volk *et al.*(2011).

4.3 Agricultural land

Thanks to its biophysical and climate conditions, Serbia has favourable natural conditions for diversified agricultural production. The soils in Serbia are extremely diverse, as a result of the great heterogeneity of the geological base, climate (nine edaphic climatic regions), vegetation and micro-fauna. Approximately 2/3 of Serbian territory consists of rolling, hilly and mountainous regions. Regarding its suitability for agricultural production, Serbia's soil potential is divided into eight fertility classes, where the first four classes represent better soils (45 per cent of total agricultural land), and classes 5-8 include the areas mainly unsuitable for tillage (and largely used as home gardens or as meadows). Two thirds of agricultural land in Serbia is in an area where crop growth is possible for over 200 days, assuming adequate and properly distributed precipitation. The climate varies between a continental climate in the north, with cold winters and hot humid summers with well distributed rainfall patterns, and an Adriatic climate in the south, with hot, dry summers and autumns, and relatively cold winters with heavy snowfall inland. An annual per capita water flow of 1.550 m^3, classifies Serbia among the water-poor areas of Europe.

However, a large part of the arable land is acidulated as a result of the uncontrolled use of chemical agents (largely a heritage of former times), and in Vojvodina it is also saline. The irrigated land is only 1.5 per cent of total arable land, whereas about 85 per cent of agricultural land is endangered by wind and water erosion. Approximately 52 per cent of Serbia's agricultural land is affected by poor drainage. Drainage systems cover 2.08 million ha. Due to lack of funds, drainage channels have been deteriorated by siltation and weed growth, and pumping stations are inactive due to a lack of maintenance. Although Serbia's irrigation system was built to cover an area of 149,000 ha, only 30,000 ha are currently irrigated, of which 90 per cent are located in the Vojvodina region; a further 47,000 ha operates only partially. Systems have ceased to function totally or efficiently as a result of lack of maintenance and a breakdown in operational management (World Bank, 2007).

The share of agricultural land in the country's total territory is 66 per cent. Agricultural land in Serbia covers about five million hectares (0.69 ha per capita), of which approximately 3.3 million hectares (65 per cent) is arable land. Compared to neighbouring countries and EU-27, Serbia has a quite favourable land per capita situation (see Figure 4.1).

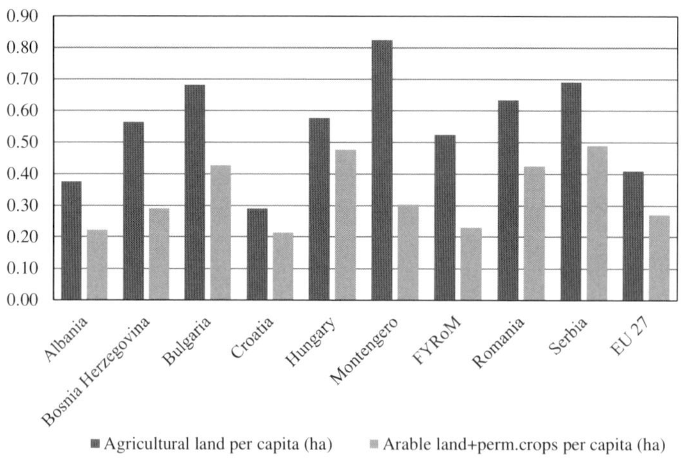

Figure 4.1. Agricultural land availability.
Source: Own calculation based on FAO database.

The most intensive agriculture is practised in the plains of Vojvodina (about 65 per cent of the country's arable land), which is characterized by stretches of flat plains and highly fertile soils. Vojvodina region is characterized by exceptionally good quality and highly fertile land. Available agricultural land per capita amounts to 1.1 ha and 10.4 ha per person employed in agriculture. These numbers are higher than in many other areas.

In Central Serbia, arable land is 66 per cent of total agricultural area, but soil fertility and climatic conditions favour fairly intensive production of high-value fruits, notably berries and vegetables. Farms are smaller; semi-subsistence farms prevail and farming systems are more diverse than in Vojvodina. The area of utilized arable land has declined significantly in the South and East Serbia region (11 per cent less hectares in 2009 than in 2000). This region stands out for the high percentage of unused land, especially in the southern municipalities, which have over 70 per cent of unused land. The unfavourable age structure of the population is the main reason for leaving land fallow, next to the lack of appropriate machinery, the poor accessibility and low quality of land, and the lack of labour. Dairy cattle dominate livestock production. Southern Serbia is mostly hilly and, therefore, farm sizes are small. Production is mostly extensive, with fruits and vegetables as main crops, whereas sheep grazing on natural pastures dominate livestock production.

In general, agricultural land use has not changed much over the past ten years. The only significant change was a decrease in the permanent crops area, which was mostly due to a decline (by almost 20 per cent) of the area under vineyards. The reason for this decline is the lack of labour. In addition, during the last decade investments in viticulture were aimed to replacing old plantations with new varieties, not the growth of area devoted to vineyards.

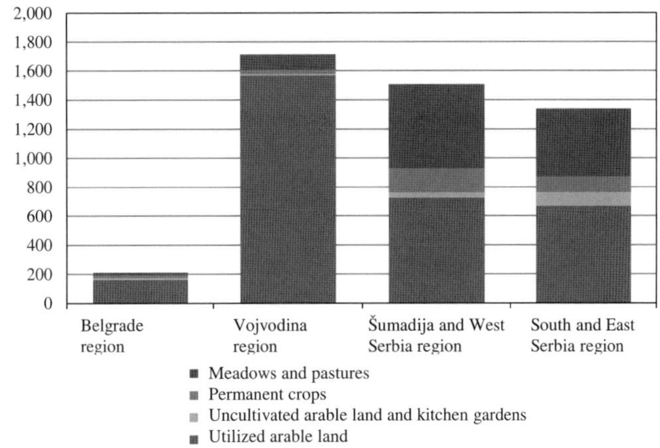

Figure 4.2. Structure of agricultural land of Serbia, by regions. Source: Own calculation based on GoRS (2011), see chapter 11.

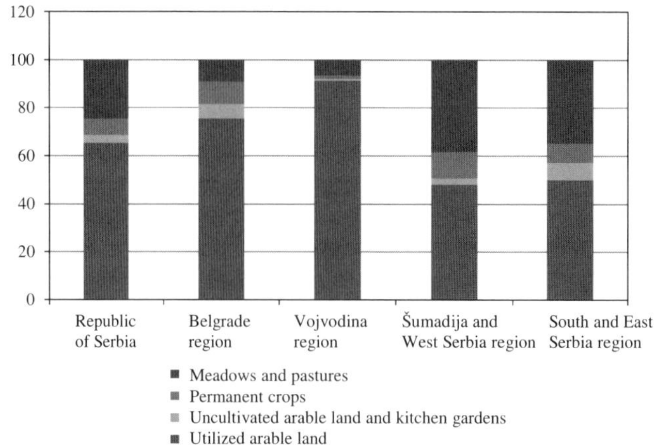

Figure 4.3. Agricultural land structure, by regions.
Source: Own calculation based on GoRS (2011), see chapter 11.

Table 4.4. Utilized agricultural land 2009, Serbia and regions, thousand ha.

	Utilized arable land			Not cultivated arable land and kitchen gardens	Permanent crops				
	Total	Sown	Fallow land		Orchards and areas under berry plants	Vineyards	Nurseries	Meadows and pastures	Total
Republic of Serbia	3,121	3,069	52	157	262	58	2	1,173	4,772
Belgrade region	161	160	1	13	17	3	0	20	212
Vojvodina region	1,566	1,563	3	11	19	10	0	110	1,716
Šumadija and West Serbia region	725	695	30	39	150	14	0	578	1,506
South and East Serbia region	669	652	17	95	76	31	1	465	1,337
Changes in agricultural land use structure (2009/2000)									
Republic of Serbia	97.4	96.9	133.9	114.8	101.4	81.5	114.8	101.5	98.9
Belgrade region	91.7	91.8	76.5	318.9	109.0	85.5	35.0	101.3	97.7
Vojvodina region	103.1	103.2	77.8	17.0	105.8	82.7	112.4	98.6	99.6
Šumadija and West Serbia region	95.6	94.2	144.5	155.6	100.7	78.4	75.1	104.1	100.0
South and East Serbia region	88.8	88.0	138.7	206.8	100.4	82.2	190.5	99.2	96.8

Note: Official data on 4.772 million hectares of UAA was reported in GoRS 2011, which differs from data on the total agricultural land by the Statistical Yearbook, which is 5.058 million ha.
Source: Own calculations based on GoRS (2011).

4.4 Farm structure

Serbia's farm structure is complex, consisting of small subsistence agricultural households, small semi-subsistence farms, large family farms, as well as privatized large enterprises with a mixed ownership structure.

Average farm sizes change slowly in Serbia. This is due to several reasons. An important factor is the slow revitalization of the rest of the economy, resulting in hidden unemployment in agriculture. Agricultural employment was also high in Romania and Bulgaria during their transition period. Most other countries, especially Czech Republic, Hungary and Estonia, had a faster transformation of their economy, which created employment opportunities and room for migration of labour from agriculture to non-agricultural sectors. This process in Serbia has been much slower, preventing the transfer and use of land, technology and capital to more efficient producers. The disruption of the labour market that took place during the 1990s in the course of the privatization process, resulted in the transfer of labour from industry to agriculture. During the 1990s agriculture became a shelter for the rural population (especially those with limited physical, human, financial and social capital) that was unable to fully participate in Serbia's changing economic and societal environment. The private sector (outside agriculture) did not have the dynamic growth that could absorb the industry workers that lost their jobs after privatization. Therefore, migration of labour from agriculture and reduction of hidden unemployment came to a halt. Hence, the agrarian structure is dominated by households with small and semi-subsistence farms, and a dual agrarian structure has been established only in the northern part of the country with a more developed and diversified economy.

According to the 2002 Census (the latest Census data available), there are about 778,900 private farms in Serbia, with an average size of 3.7 ha. Furthermore, land is fragmented in 4 plots per farm, on average. According to the 2002 Census, over 75 per cent of private farms comprise less than 5 hectares and only about 6 per cent exceed 10 hectares (see Figure 4.4.). It is estimated that private family farmers own approximately 82 per cent of the 5.1 million hectares of agricultural land.

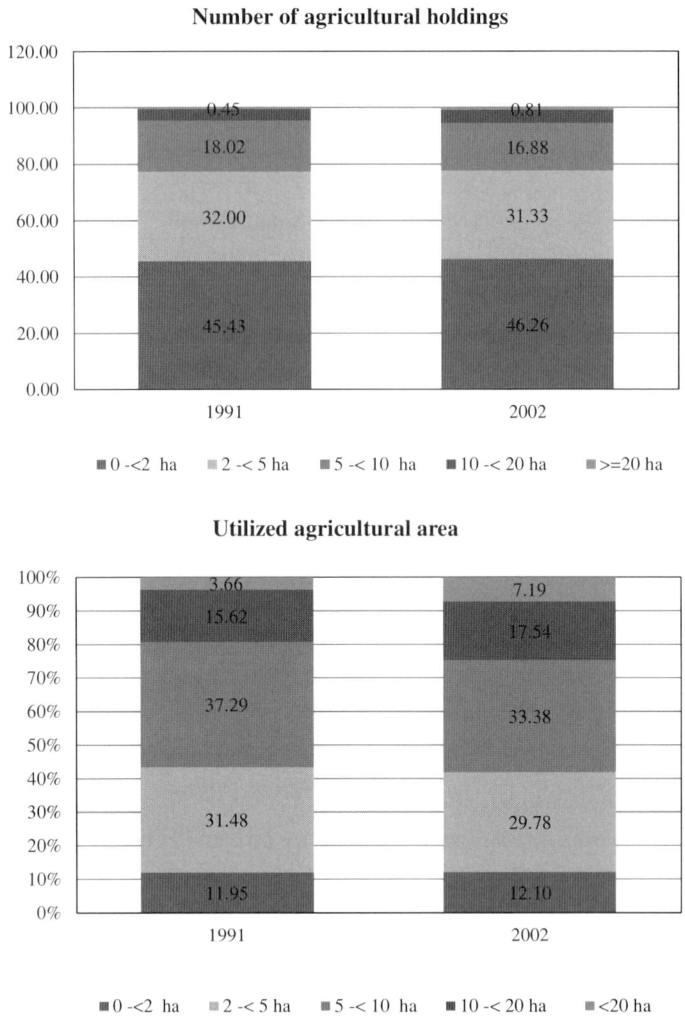

Figure 4.4. Number of private family farms and utilized areas, according to Census 1991 and 2002.
Source: Census of Population, Households and Dwellings (2002).
Note: this structure refers only to arable land of family farms. Meadows are not included, nor are state land and land of big agricultural enterprises.

Up to now a trend towards larger farms has been visible only in the northern part of the country, especially in areas with a pronounced trend of an aging rural population. The land market in Vojvodina is operational, yet leasing prevails over land purchases. This region is also eminent for the rather large number of small farms -- farms of a size up to 3 ha cover 66 per cent of the total number of farms in the region (see Figure 4.5). This high percentage of small farms can be explained by the fact that because of the relatively favourable development of infrastructure and the adequate level of social services in this rural region, a considerable number of households use the so-called 'farms' merely for residential purposes, though in statistical terms these same 'farms' are defined as normal farms. The number of farms with 3-10 ha is significantly lower than in other rural regions and this is evidently a consequence of the divergence of farms according to their sources of income. On the other hand, farms of 10 ha and more cover as much as 37 per cent of the agricultural area in this region. Also, 30 per cent of farms in Serbia of over 10 ha are situated in this region.

The farm structure in the central part of the country (Sumadija and West Serbia region), with a larger agricultural population and a mixed farming economy is dominated by 'middle-sized' farms with 3-10 ha per farm (Figure 4.5). A significant difference in comparison to other rural regions is a very small share of farms of a size over 10 ha in the total number of farms.

In South-East Serbia 60 per cent of farms have less than 3 hectares. According to the law on the maximum size of farm holdings in the Communist era, farms in the mountainous parts of this region had the right to an area double the size allowed in the plains. Due to the high migration rate to the towns and abroad, the farms were not fragmented, as it was the case in regions with more favourable development performances. In the southern part of the region, the population age structure and dynamic emigration influenced the demographic movement and farm regrouping according to the property size.

The Serbian agriculture is dominated by small and semi-subsistence farms with little produce going to market. Cvejic *et al.* (2010) estimate that about 60 per cent of households in Serbia do not have surplus or sell only occasionally. These farms generate low yields and lack specialization. These farms survive because of social reasons, as although they are not economical, these farms provide food security for more than half of these households. Analyses of changes in farm structure and land ownership fail because of a lack of reliable data. According to Cvejic *et al.* (2010) the average size of agricultural land which is cultivated by commercially oriented households is about 10 ha in Central Serbia and about 30ha in Vojvodina. These data indicate average farm sizes that are several times greater than the Census 2002 results and testify to the much more active land market than the official statistics registers show.

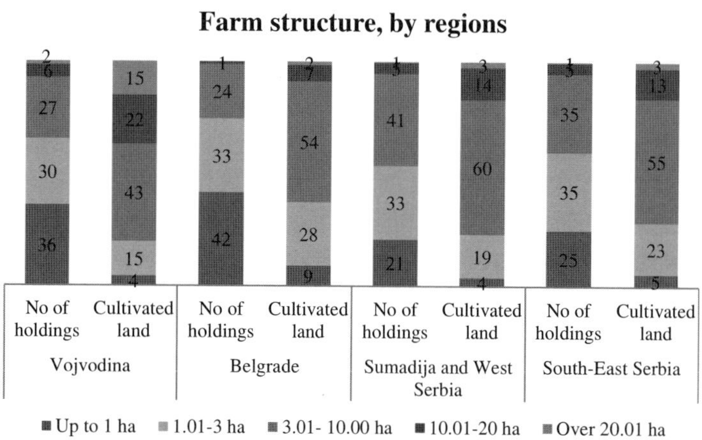

Figure 4.5. Farm structure according to Census 2002, by regions. Source: GoRS (2011).

In addition to the privately owned land, land is owned by other entities, like cooperatives and enterprises. Roughly 10 (in Central Serbia) to 25 per cent of land owned is owned by agricultural enterprises and cooperatives, the rest is owned or rented by family farms. The structure of land owned by enterprises and cooperatives is quite different in these two regions. In Vojvodina, enterprises and cooperatives possess the arable land and meadows, while in Central Serbia they own much more pasture land.

The enterprises currently active are privatized. The most common challenge they face is to raise operational capital to become more productive and pull in investment capital for reinvestment in established fixed assets. Many of these enterprises have under- or unutilized assets such as buildings. There are, though, large scale enterprises using modern hybrid production systems and having levels of efficiency similar to those in the EU.

Table 4.5. The structure of commercial farms in 2008.

Type of production/ farm size	Total	Without land	Under 50 ha	51-100 ha	100-500 ha	501-1,000 ha	1001-2,500 ha	Over 2,500 ha
Number of agricultural businesses	812	158	136	71	184	79	120	64

Source: SORS Statistical yearbook 2009.

Box 4.1. Land privatization process.

During the Communist period agricultural land was confiscated without compensation and farms were run in a cooperative or state ownership structure. Land privatisation began in the early 1990s, effectively with the application of the 'Law on the manner and conditions of recognition of rights and restitution of land,' etc.[1] However, former owners or their successors were only compensated up to a maximum of 10 ha (15 ha with forest area). The result was that a substantial area of agricultural land remained state owned, and was farmed by agricultural vertically integrated companies and cooperatives. Since the Parliament adopted the "Law on registration of confiscated property" (Official gazette of RS, no. 45/2005) a start has been made to return the confiscated property and to compensate former owners. Agricultural enterprises and individuals, however, have great difficulties in claiming land as the land registry and cadastre are typically out-dated and/or incomplete. Meanwhile, the state applies the principle that all land of which companies are unable to prove they are the owners shall be considered as state property. This privatization process caused great controversy and turbulent public reactions. The lack of clear ownership rights for a significant part of the land is a hindrance to the proper operation of the land market, although land tenure in Serbia is overwhelmingly private. However, today the majority of public property, which originates from land confiscated from former owners, remains in state ownership, although it can be rented out. Currently state owned land amounts 900,000 ha approximately, of which around half of the area is leased. Most of these land (300,000 ha) is located in Vojvodina.

[1] 'Law on the manner and conditions of recognition of rights and restitution of land that was transferred to public property from agricultural land fund and the confiscation of the outstanding commitments from the compulsory purchase of agricultural products' ('Official Gazette' No. 18/91).

4.5 Production: size and regional diversity

By the early 1980s Serbia experienced a significant growth of agricultural production (3.5-4 per cent annually), which stagnated in the late 1980s and declined sharply in the 1990s. Extremely unfavourable production and economic indicators regarding Serbian agriculture in 1990s are the marked fluctuation and negative trend in production of almost all agricultural products; these reflect low labour productivity and a low level of marketed production.

The period 2000-2010 was characterized by substantial annual fluctuation of agricultural production, which generally remained lower than in the pre-transition period. Agricultural production in Serbia, which is of a typically extensive nature, is also strongly influenced by the weather conditions, especially droughts.

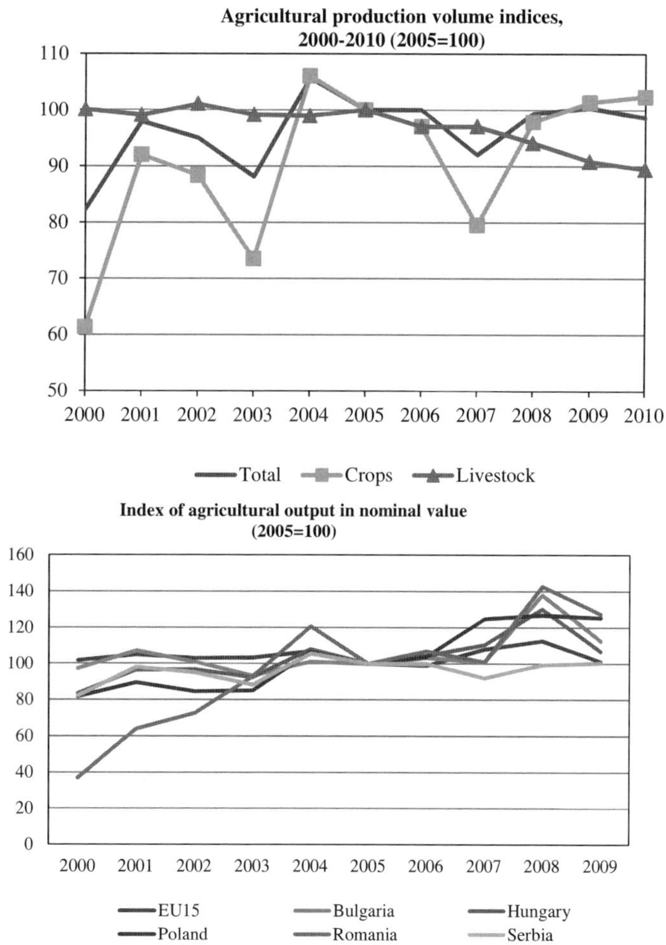

Figure 4.7. Agricultural production in volume indices (a) and value indices (b). Final (net) production only. Source: SORS.

Crop production is a high and increasing share in Serbia's agricultural production, with maize, wheat and fruit as major crops (see Table 4.6). Livestock's share in the sector's total agricultural production value declined, largely because of developments in the meat sector, where the contribution of both pig meat and beef production shows a downward tendency in recent years. The contribution of milk production to the agricultural output value fluctuates between 7.5 and 9 per cent.

Developments in the structure of gross agricultural output (GAO) in several countries that entered the EU in the 2000s show a diverse picture (Figure 4.8). In fact, the share of livestock production has declined after the accession, most markedly in Bulgaria, Romania and Hungary while the livestock sector in Poland and Slovenia kept its position.

Table 4.6. Share of certain agricultural products in the value of total agricultural production, 2002-2009.

	2002	2003	2004	2005	2006	2007	2008	2009
Wheat	7.6	5	8.1	5.8	5.6	6.6	7.1	6.5
Maize	22.7	17.7	24.2	25.9	22.6	17.2	25.1	26.5
Sugar beet	1.9	1.6	2	2.2	2.4	2.5	1.6	1.8
Sunflower	1.5	2	2	1.6	1.8	1.8	2.6	2
Potatoes	3.7	2.7	3.1	3.2	1.1	2.9	3.1	3
Fruit	1.3	3.9	4.5	3.6	5	6.7	5.9	6.4
Grapes	4.6	3.4	2.5	1.3	2	2.4	2.2	2.4
Beef	6.4	7.7	6.1	7	7.2	7.6	6.3	5.8
Cow's milk	7.8	9.1	7.3	8.2	8.3	9.2	8.3	8
Pig meat	15.4	15.8	12.1	14.3	13.7	14	10.8	11
Poultry	2.9	2.5	2.2	2.4	2.6	2.6	2.4	2.5

Source: SORS.

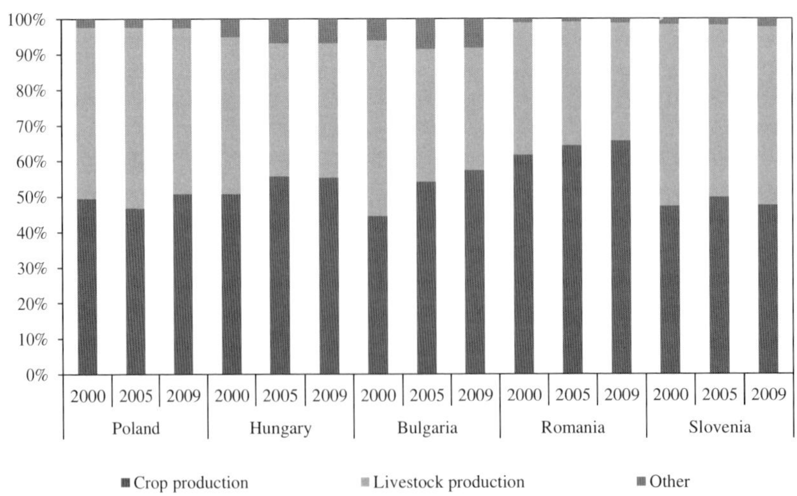

Figure 4.8. Structure of GAO, before and after accession.
Source: Eurostat.

4.5.1 Crop production

Cereals (maize and wheat) dominate the agricultural production structure. The area under cereals accounts for about 60 per cent of arable land, with a decreasing trend over the past years (see Figure 4.9). The area under cereals is declining as the production of wheat was extremely uncompetitive compared to other crops. However, cereal production still accounts for about 30 per cent of total value of the agricultural production in Serbia.

Technical progress -- through the use of mechanization, better seeds and so on -- that resulted in growth of yields and production came to a halt in the early 1990s. When trade sanctions ended, and markets opened, the economic system stabilized in early 2000s, and yields in agriculture began to rise again. However, average yields have still not reached the level they had in the pre-transition period, and are significantly below those in the neighbouring countries like Croatia and Hungary. The reasons for low yields are manifold, such as technological backwardness, relatively low and fluctuating prices, lack of credit and an inefficient system of knowledge transfer in agricultural practices. Recent studies have shown that the use of uncertified seed is about 50 per cent, and that the use of mineral fertilizers decreased by half compared to the technical optimum (Economic Institute, 2010).

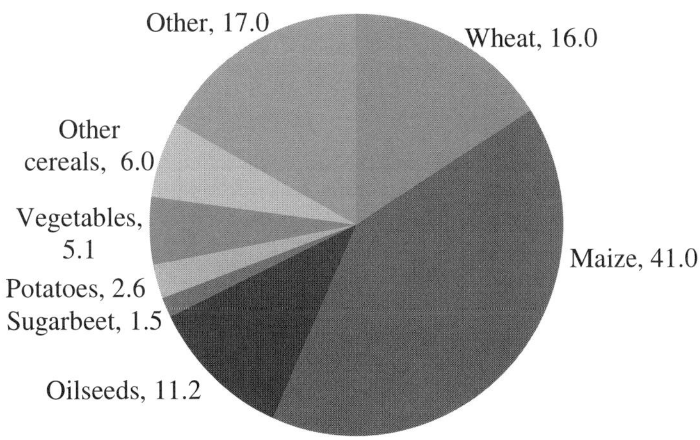

Figure 4.9. Breakdown of harvested area by main crops (2009).
Source: Statistical Office of the Republic of Serbia.

Maize is Serbia's biggest single commodity with an average production of 5.6 million tonnes during last decade, which was produced on approximately 1.2 million ha. Over the last 5 years maize production is increasing. In 2010 it reached a record level of 7.2 million tonnes with an average yield of 6 tonnes (see Figure 4.10). Maize is the most important export product of Serbian agriculture. Maize is the major crop in Serbia that producers can easily store on their farms. Farmers harvest the crop in October and November and it can either be stored on farms to be naturally dried or taken to artificially drying facilities. About 80 per cent of total consumption is used for animal feed, while the rest is used for human consumption, alcohol and starch production. There are two main starch production factories currently operating in Serbia. Currently there is one bio-ethanol plant built in Vojvodina.

The latest technology has been used for building this factory, with a maximum capacity of 100,000 tonnes per year. This factory is able to produce biodiesel of standard EU quality (EN 14214). Due to the lack of government regulation on bio-ethanol production, this factory is still not producing bio-ethanol but is used to provide crushing capacity for sunflower and soybeans. Once the plant starts producing bio-ethanol, maize production in Serbia can drastically increase.

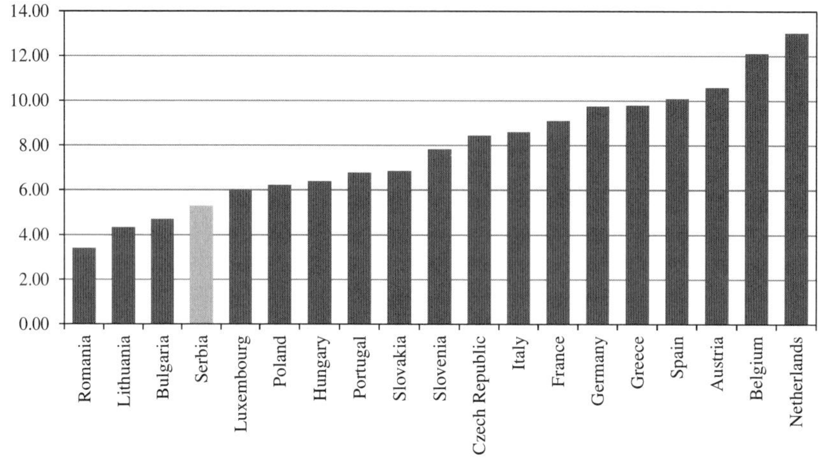

Figure 4.10. Maize yields in EU countries and in Serbia (2009).
Source: Eurostat, SORS.

Wheat is the second biggest commodity in the Serbian cereals sector. Between 2000 and 2010 Serbia produced some 2.3 million tonnes of wheat annually on approximately 590,000 ha. Over the last 5 years the area used for wheat production varied, but generally decreased. The sown area in 2010 was lower than ever -- 480 thousand hectares. Average yield of wheat over the last 5 years has been 3.6 t/ha while the EU25 has had an average yield of approximately 5.5 t/ha (see Figure 4.11). Although Serbia is not competitive compared to major Central European exporters such as Hungary, Serbia's cereal production is contributing significantly to providing a sufficient supply for Western Balkan countries. Total domestic wheat consumption in Serbia is estimated at about 1.5 million tonnes annually. Wheat for human consumption is estimated at 1.4 million tonnes annually and per capita consumption is estimated at 185 kilos, which is significantly higher than consumption levels in most European countries.

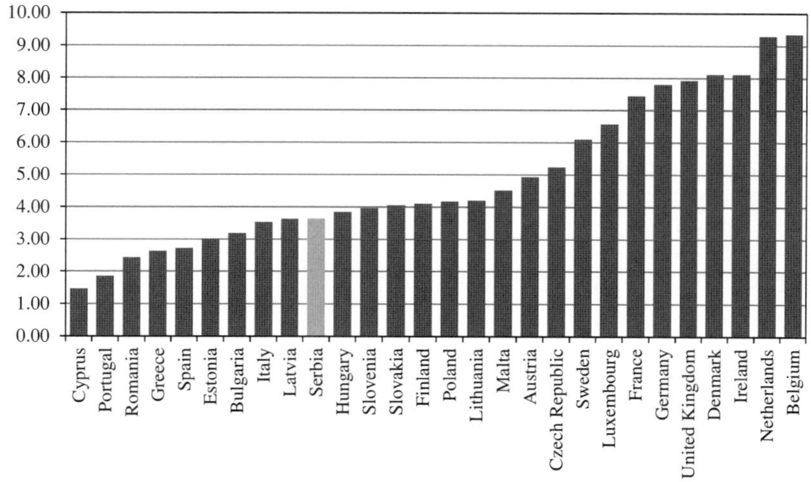

Figure 4.11. Wheat yields in EU countries and in Serbia (2009).
Source: Eurostat, SORS.

Fruit and vegetable production accounts for about 12 per cent of the value of agricultural output, and positive trends were recorded in the past years. Fruits and vegetables occupy about 60 per cent of agricultural land and are predominantly cultivated on private holdings in Central Serbia (about 99 per cent). In this sector significant progress has been made in the improvement of standards in primary production and processing and in the linkages between them. Serbia has good climatic conditions for growing many varieties of fruits. Over the last years, thanks to favourable credit conditions for purchasing irrigation systems and building green houses, production has been significantly intensified. Demand for Serbian vegetables and especially fruits is growing, with exports expanding not just in quantity, but also in the number of countries to which Serbia is exporting.

Tree fruits bring large yields, although the Serbian apple yields are far less than that which top-producers in the EU achieve (see Figure 4.12). Apples, plums, sour cherries and raspberries are major fruits in Serbia, while fruits with the potential for rapid and sustainable growth are peaches, apricots, and sweet cherries. Because of the influence of the weather on agricultural production, quantities produced vary from year to year, affecting the bulk purchase price of the products. Fruits are grown on approximately 330,000 ha (tree fruits, strawberries, raspberries, blackberries). Fruits and especially berry fruits contribute significantly to Serbian agricultural exports. Apart from being sold fresh, Serbian fruits are also often processed. Plums are used in the production of šljivovica, a traditional Serbian plum brandy. With an overall production of around 80,000 tonnes/year, Serbia has long been one of the major raspberry producers in the world and still holds a leading position. Most fruit in Serbia is grown on small family farms.

Major vegetable crops produced in Serbia are potato, tomato, pea, cabbage and kale, onion, pepper, various varieties of beans, melon and watermelon, carrot, and cucumber.

Viticulture is widespread over the whole country with an average production of 380,000 tonnes of table grapes and wine grapes. Most of the grapes are used for wine production (annual production 1.85 million HL). Over the last five years the planted area increased slightly and so did the physical volume of the production.

The industrial crop production (oilseeds and sugar beet) accounts for 7 per cent of the total value of agricultural production of Serbia. Over the past few years, the sector has significantly recovered compared to the 1990s. Area sown with industrial crops has recorded permanent growth since 2000. Its share in arable land increased in the period 2000-2010 by about 16 per cent. Opening of the foreign market, budgetary support, export subsidies and privatization of the processing capacities contributed to fast revitalization of industrial crop production. More than in other segments of the food chain in Serbia, this sector has set up a trade chain with a positive impact on the growth of the area cultivated and of total production.

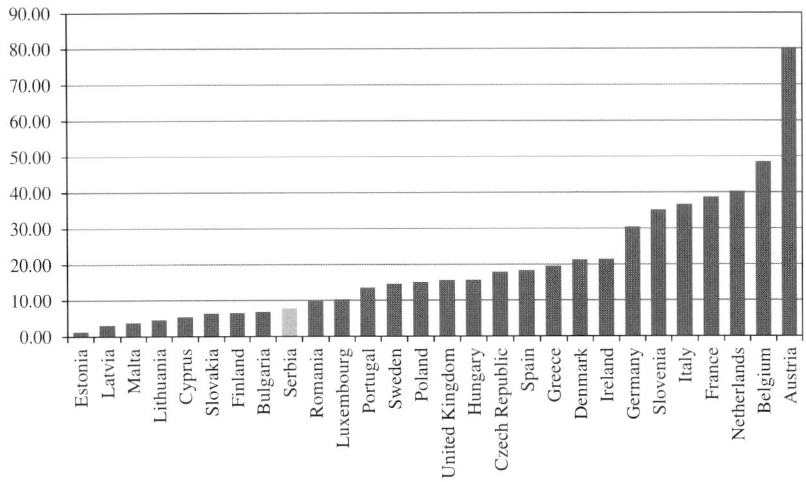

Figure 4.12. Apple yields in EU and Serbia (2009).
Source: Eurostat, SORS.

The largest part of the production of industrial crops is concentrated in Vojvodina. Sunflower and soybean yields in this region exceed 2t/ha and are among the highest in the Balkan region. Sugar beet with an overall cultivation area of 65,000 hectare and production of 3 million tonnes is another important Serbian crop. Sunflower cultivation extend over 170,000 hectares, yielding a production of 370,000 tonnes/year. The soybean yields are among the highest in Europe (over 2.5t/ha). Producers use domestically produced seed varieties that have a high genetic potential, and provides a grain of high quality. Serbia is a GMO free country, which makes soybeans from Serbia/Vojvodina competitive on the international markets (Economic Institute 2010).

4.5.2 Livestock production

The livestock sectors make a significant contribution to Gross Agricultural Output (GAO), currently estimated at about 30 per cent (see Table 4.6). It is even higher if own production traded and consumed locally would be taken into account as well.

Despite its importance to the rural and domestic economy, the number of all livestock has been decreasing over a last two decades, with the strongest decline in the years 2000-2002. This has had important effects on the production levels in the livestock sector. The sheep flock shrank significantly in the mid-1990s, with the result that sheep milk production fell by 35 per cent in that decade compared to the 1980s (see Figure 4.13). Since 2000, the sheep flock shows a further decline of about 150,000 animals to 1.647 million head in 2009. All in all, the sheep milk production is only 40 per cent of the level it used to be in the pre-transition period. During the period 2000-2009 the number of cattle decreased from 1.25 million head to 1.00 million head. The decline particularly occurred in the milking cow herd, of which the numbers fell from 750,000 animals in 2000 to 550,000 in 2009. Despite the smaller herd, however, the cow milk production increased compared to the average level of production in the 1990s, reaching the pre-transition production level again. This is due to increasing yields per cow.

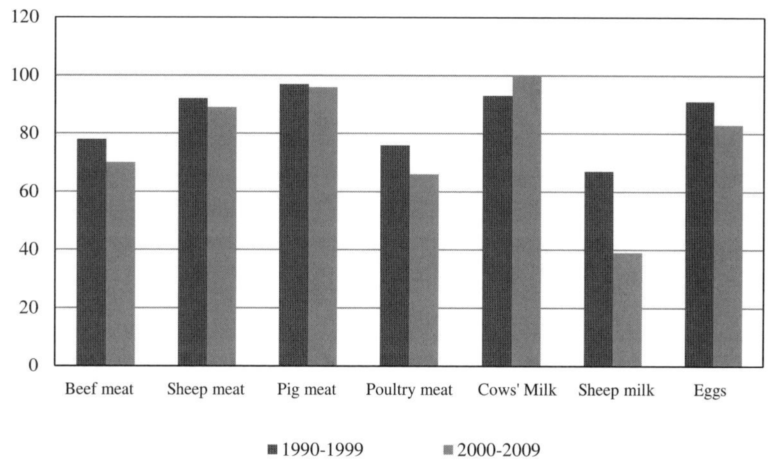

Figure 4.13. Changes in production of meat, milk and eggs (1980-1989=100).
Source: SORS

Still, the average production per cow is far below those in most EU countries (see Figure 4.14). Pig and poultry numbers fell too, which had a stronger impact on poultry production than on the pig meat production (see Figure 4.13).

The main reasons for the reduction of livestock production were the decline in consumption, fluctuating feed prices and a limited external market. In recent years there have been substantial efforts by the state to revitalize livestock production (see Chapter 7 on policies). However, next to the unfavourable economic situation for cattle breeding, the privatization process resulted in a further reduction of the number of cattle. Many new owners of (the larger) agricultural enterprises seek to get rid of livestock and to concentrate primarily on crop production.

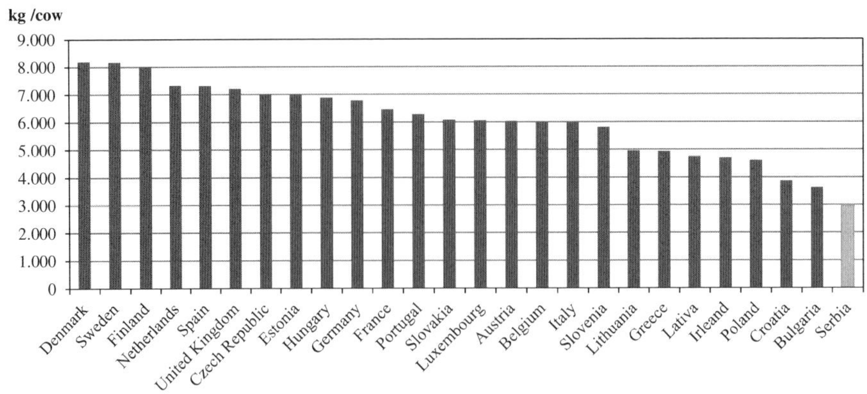

Figure 4.14. Milk yields in Serbia and the EU, 2009 (thousand litres/cow). Source: SORS

Small scale producers are prevailing. They traditionally work with low-intensity farming systems characterized as being semi-subsistence and composed of locally adapted breeds. Their existence is important from the perspective of preserving the environment and rural economy. Small scale producers are mostly linked to dairy and meat production. There are also some differences in the importance of family holdings in production. The percentage of animals reared on family farms is higher in central Serbia, except for beef cattle.

The regional differences in trends and production systems for livestock production are substantial. In Vojvodina and the low lands there is a long tradition of crop production while livestock is often less important. In the hills and mountain areas of Serbia livestock production is the most important source of income for family farms. Within this production system, there is also a strong regional differentiation. A common order in the relative importance of certain species of livestock is cattle-sheep-poultry-pig. In Vojvodina and flat areas the order is reversed. In those regions cereal production and technological progress has changed the order to pig-poultry-cattle-sheep.

Cattle breeding is the most important agricultural activity in Serbia, primarily for the small and medium sized family farms. It is also very much an integrated system of both milk and beef production. Serbia has a long tradition of cattle breeding and producers' husbandry skills are high. The structure of the national herd significantly improved since 1970's[1]. In the past it was dominated by the autochthonous breeds (Busa, Podolian), but was nevertheless regarded as being in a strong position with good potential for both milk and beef production. Nowadays, the dominant breed is the Simmental cattle. The current structure of the national cattle herd in Serbia is as follows: 35 per cent of pure bred

[1] Current cattle breeding programs for Simmental and Holstein cattle have their origins in the 1970's, evolving from SFRY State Programs.

Simmental, 50 per cent of domestic spotted cattle of the Simmental type or the domestic Simmental breed, 5 per cent black and white cattle with various percentages of HF genes, as well as pure Holstein-Friesians, and 10 per cent of primitive breeds and cross-bred cattle. Farms of less than 3 hectares have a one cow, on average, but represent about 25 per cent of the total herd size (see Table 4.7). Most animals entering supply chains come from herds of an average size of no more than 5-8 animals (Gill *et al.*, 2010).

Table 4.7. Number of holdings (x1,000), cattle (x 1,000), and cattle numbers per size category.

Type of production/ farm size	Total	Without Land	Under 1 ha	1.01-3 ha	3.01-5 ha	5.01-8 ha	8.1-15 ha	Over 15 ha
Number of agricultural holdings	778.9	6.3	208.1	254.8	135.2	96.8	62.3	15.3
No of cattle	748.3	5.9	28.4	141.5	161.1	177.9	170.0	63.4
Average number of cattle/holding		0.94	0.14	0.56	1.19	1.84	2.72	4.1
% of flock/herd by holding type								
Cattle	100	1	4	19	22	24	23	8

Source: Census (2002).

The beef aspect is almost totally reliant on maintaining and developing a competitive dairy industry, and is vulnerable to the adoption of EU legislation if this encourages small-scale milk producers to cease production. When the dairy sector becomes more efficient and increases productivity, it will affect the dairy sector in that more animals are selected for breeding. The number of cattle in the last decade has declined by 20 per cent, although the growth of milk production and genetic improvement were minimal. This has largely been a result of an unstable support, the lack of an organized domestic market and fluctuations in feed prices.

Milk production in Serbia has been rather stable at around 1.5 million tonnes since 2000, with a small decrease through a 3.7 per cent drop from 2008 to 2009 in response to a noticeable decline in the milk price. According to the Census 2002 there are approximately 518,000 cows on 269,000 farms, giving an average herd size of 1.9 cows. Herds are larger in Vojvodina (3 cows on average) compared to those in Central Serbia (1.8). From regional standpoint yields are much higher in Vojvodina (3,890 litres/cow) than in Central Serbia (2,730). The difference comes from the breed structure: 52 per cent of cows in Vojvodina are Freisian-Holstein compared with less than 8 per cent in Central Serbia where Simmental and Simmental-cross cattle predominate.

According to ALC (2010) about 250,000 farms with herds of 1-5 cows account for 95 per cent of the herd, 89 per cent of all milking cows, 68 per cent of milk production and 59 per cent of all milk delivered to dairies. This production contributes significantly to the living standard and cash income of family farms. Herds of more than 50 cows account for less than 1 per cent of herds, 5 per cent of cows, 11 per cent of milk and 14 per cent of milk delivered to dairies. Most of the production in this group of large producers comes from corporate farms with over 200 cows. The group of 'emerging family farms' -- those with 20-50 cows and potential for growth -- is still very small, numbering just under 500 farms.

The milk market in Serbia consists of the formal market, which relies on dairy plants that process milk and sells final products through supermarkets, and the informal market, where milk is sold directly to local consumers. Informal markets also include farm processed products such as white cheese and kajmak, and the products sold at farmers' markets or directly. According to SORS, 65 per cent of milk for human consumption is delivered to dairies and passes through the formal market. The rest is informal processing and selling. Farmers' markets are a very important income source for dairy farmers and an important outlet to a few million consumers.

Sheep production in Serbia has a long tradition and is in harmony with natural conditions. The main product of sheep sector is the meat, as well as both milk and wool. Their contribution to households income is, however, very important, although figures are misleading as only a small percentage enters the registered food supply chain. In hilly and mountain areas milk and cheese are still the most important household and commercial products. The dominant breeds are the Tsigaia and the Pramenka. Besides there are about 20 local breeds recognized throughout the Balkans that are regionalized. A high concentration of the sheep herd is on hill and mountainous areas (300m above sea level), especially in the east and south of Serbia. Production is dominated by extensive grazing land that is not suitable for other types of agricultural production. A small number of sheep are kept in lowland areas. Sheep numbers were most affected in the mid 1990's. After a steep decline in the 1990s, the national sheep flock has reportedly remained relatively stable from 2000 onwards. The sheep meat production is outside of commercial markets as a result of on-farm sale of the animals and meat (directly to the consumer); only some 20 per cent of total sheep-related production is in the form of milk sold to dairies.

The commercial pig industry in Serbia has long used a pyramid organization of production of breeding animals and fatteners. Pig breeding companies are based on nucleus herds of top quality pedigree pigs. Pig production in Serbia is concentrated in large herds; an estimated 80 per cent of pigs are kept in herd sizes of over 1000 head. Larger commercial production is reported to account for an estimated maximum of 45 per cent of the total pig production with the remainder of the national herd kept in smaller units on smaller family farms. The most popular breeds in Serbia are Landrace and Large White (Yorkshire). Pig meat production has decreased since 2000. The main reasons are the fluctuating feed prices and competition from imported pig meat. The Serbian processing plants are unable to export pig meat and by-products due to the vaccination programme that is in progress and which should lead to eradication of classical swine fever.

In the last decade the total national poultry flock had inverted trends compared to other livestock sectors. Although there was a drop (probably affected by avian flu scares) around 2005-2006, production has continued to steadily increase up until now. Commercial poultry chicken (broiler) production in Serbia is composed of a few large companies, with modern slaughterhouses, and thousands of small-scale farms. The production of small farms, or even backyard holdings, accounts for a large percentage of total production. Many of these small farms produce poultry for their own consumption, occasionally for sales outside legal market channels. The production of poultry meat in organized channels is dominated by a relatively small number of producers and processors. However, poultry meat plays a key role in the meat sector and has increased its share in consumption.

4.6 Prices

Prices of agricultural and food products during 1990s were affected by the pressure of inflation, monetary disturbances, and frequent changes in their administrative control and support measures. Agricultural prices were influenced also by the monopoly position of the domestic processors as a result of the economic isolation of the country, high import protection, the presence of grey economy, and compensatory payments that were common in the circumstances of hyperinflation.

After the political reforms in 2001 inflation rates went down and the exchange rate stabilised. These developments consequently led to more stable prices in the agricultural markets, and the terms of trade for agriculture improved. Government policy reforms opted for liberalisation of the agricultural markets, but in practice market interventions remained. The government intervened on several occasions (in the years affected by natural disasters) with emergency purchases of wheat by the Directorate for Commodity Reserves, along with the banning of exports of cereals and industrial crops. Emergency imports of vegetable oil, flour and meat was also undertaken in cases where it was estimated that the prices in the domestic market had been inflated beyond a reasonable level due to the monopolistic position of processors.

Revitalization of price policy and the mechanisms of support to agriculture started in late 2000. The Federal Government (FRY) controlled the guaranteed prices, levies and export incentives, which were the most important instruments of agricultural price policy of the country (up to 2002). At that time the interests of the constituent republics within the federation (Serbia and Montenegro) in the field of agriculture were completely opposed and conflicting. During 2001-2003 the Serbian government choose a system of orientation prices (which were not precisely defined) to purchase wheat for the Directorate for Commodity Reserves. The agreed price of industrial crops (in 2002 and 2003) were used for the budgetary incentives to producers (defined per ha sown area and yield) and were paid by processors, based on signed contracts for production and purchase.

The market of agricultural inputs was liberalized at the beginning of transition by the new customs law. In addition, until 2004 the government intervened in the input market by selling fertilizers and animal feed from the state reserves at favourable prices and through donations .

During 2000s agricultural price policy was based on a combination of producer and input subsidies, and high levels of tariff protection (see chapter 7). Price policy reform was initiated at the beginning of 2000s, in conjunction with the reform of trade policy required for WTO membership. Premiums for the most important crops and milk, as well as input subsidies, were essential agricultural support instruments. The system of support for livestock production also included headage payments, as well as support to selected livestock breeding programmes and to the management of genetic resources. In general, market price policy was not significantly transformed since 2000 -- the applied mechanisms were ad hoc solutions. Lack of a long-term programme and stable budgetary resources, institutional disorder and inconsistency of support measures caused rather heavy price fluctuations and constantly changing parities. In addition, unfavourable weather conditions (2002, 2003 and 2007) disturbed market balances.

The lowest recorded level of prices (in euro) in the period since 2000 was in 2005, after which prices have been rising again. The only noticeable pattern in prices during the period 2000-2008 was that the prices of sugar beet and livestock were more stable compared to the other agricultural commodities.

Figures 4.15 and 4.16 below present price developments of several agricultural commodities in Serbia compared with those in the EU and neighbouring countries.[1] The

[1] Prices for Serbia are calculated as weighted unit values from data on quantities and values of purchased products.

credibility of the comparison of these producer prices is limited because of exchange rate problems and probably different methods of price monitoring (Volk *et al.*, 2010). In spite of all limitations, this overview shows that the Serbian producers of the commodities presented here are relatively price competitive compared to those in the other countries taken into account.

Serbia is among the European countries with the lowest prices of agricultural products. Price competitiveness is reflected favourably on the growth of exports, but on the other hand unstable and low prices of agricultural products slow down the technical progress and generate rural poverty.

Maize prices (USD/t)

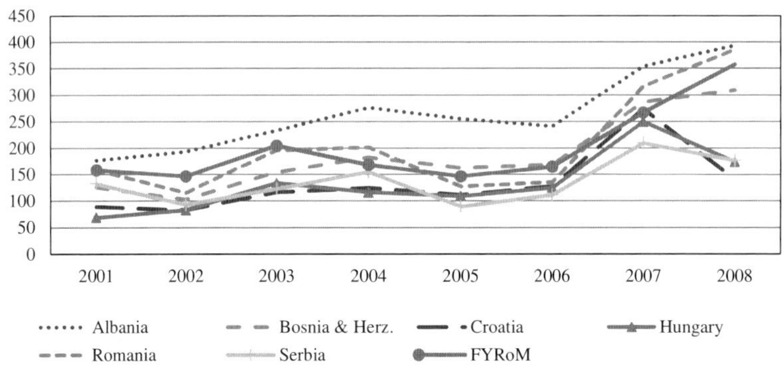

Sugar beet prices (USD/t)

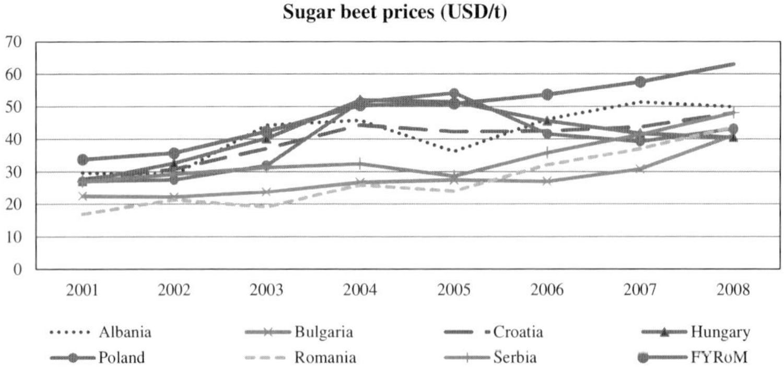

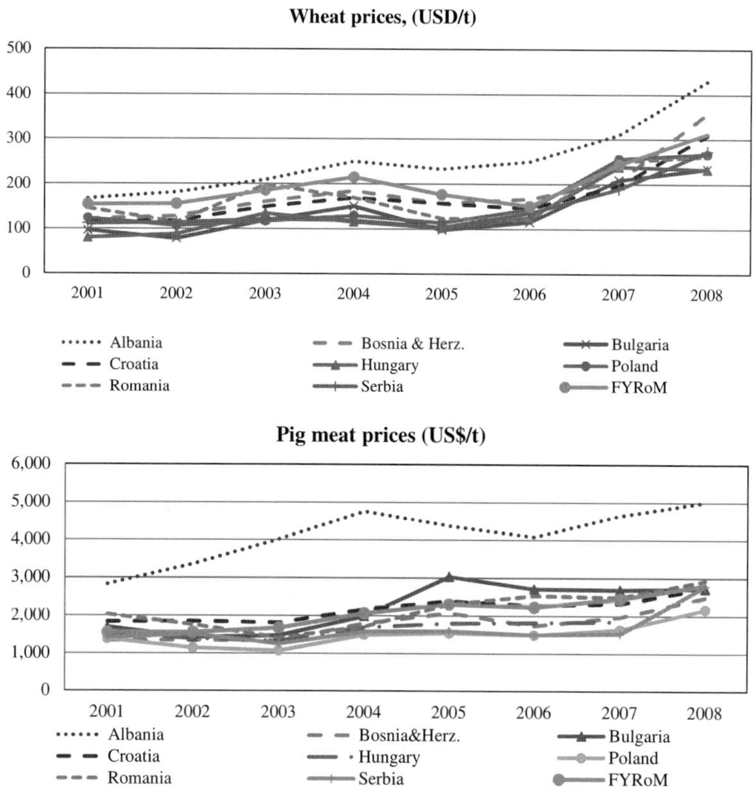

Figure 4.15. Price levels of agricultural commodities in Serbia and neighbouring countries.
Source: SORS and Faostat database.

Wheat producers prices (euro/t)

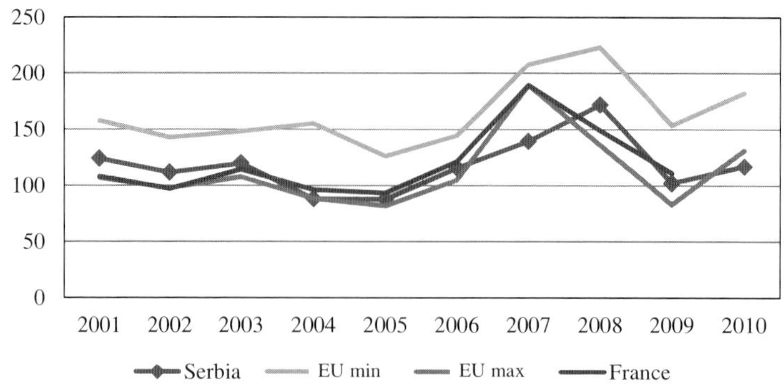

Maize producers prices (euro/t)

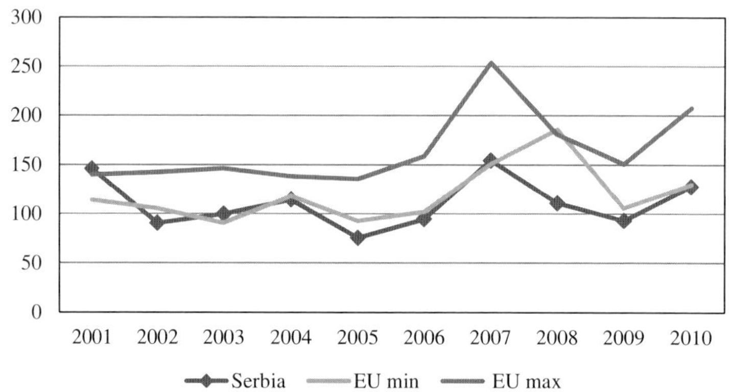

Sugar Beet producers prices (euro/t)

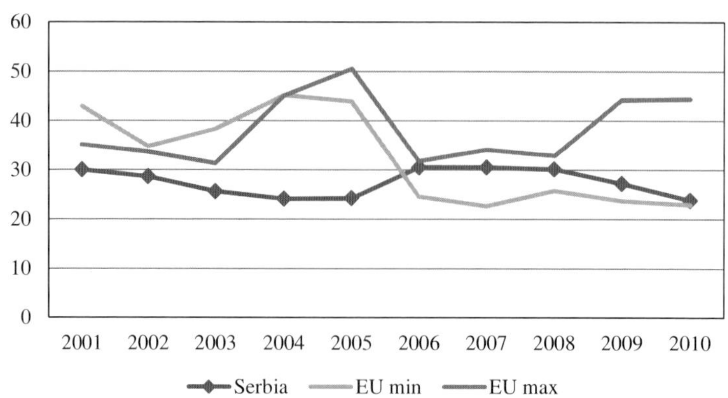

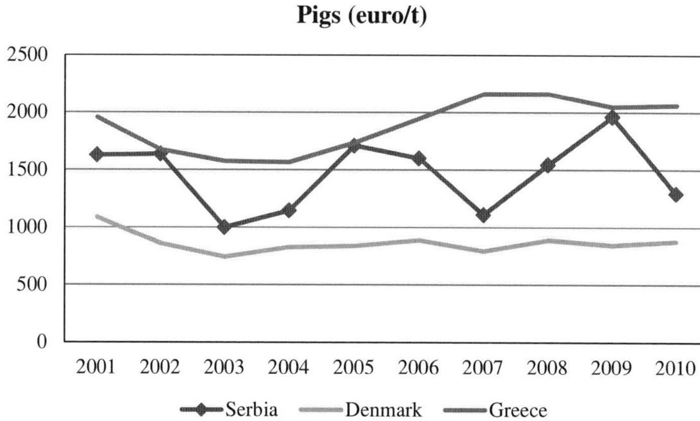

Figure 4.16. Price levels of agricultural commodities in Serbia and EU countries. Source: SORS Faostat database.

There are no official statistics of agricultural input prices in Serbia. Bogdanov *et al.* (2010), however, show that the growth of input prices (i.c. fertilizers and urea) during 2005-2009 was faster than the growth of the agricultural product prices, and that despite the nominal price increase, the economic returns of the agricultural producers remained relatively unfavourable (with the exception of 2007-2008). Table 4.8 shows gross margins (returns minus variable costs) of several main crops in Serbia. The figures show strong fluctuation from year to year, yet also indicate that over the years sugar beet has the highest gross margins.

Table 4.8. Gross margin of the main crops in Serbia (dinars/ha).

	2005	2006	2007	2008	2009
Wheat	134	149	182	429	-49
Maize	211	371	315	2	51
Sunflower	28	83	417	355	45
Soybean	184	216	379	398	217
Sugar beet	485	626	625	759	609

Source: Bogdanov *et al.* (2010).

4.7 Labour and land productivity

Data on labour productivity in the Serbian primary agricultural sector are particularly scarce. Information from sources like the Statistical Office of Serbia, the Labour Force Survey and the World Bank are inconsistent due to differences in methodology, time frame and coverage of the sector. Yet, most importantly, there is a lack of data on returns and costs of production. Therefore, it is very difficult to obtain a clear picture of farmers' income. Using Gross Value Added (GVA) as a proxy for income, it appears that agricultural income varies significantly from year to year. Gross value added per person employed and per ha is lower than in neighbouring EU countries and close to the levels of Bulgaria and Romania (see Figure 4.17).

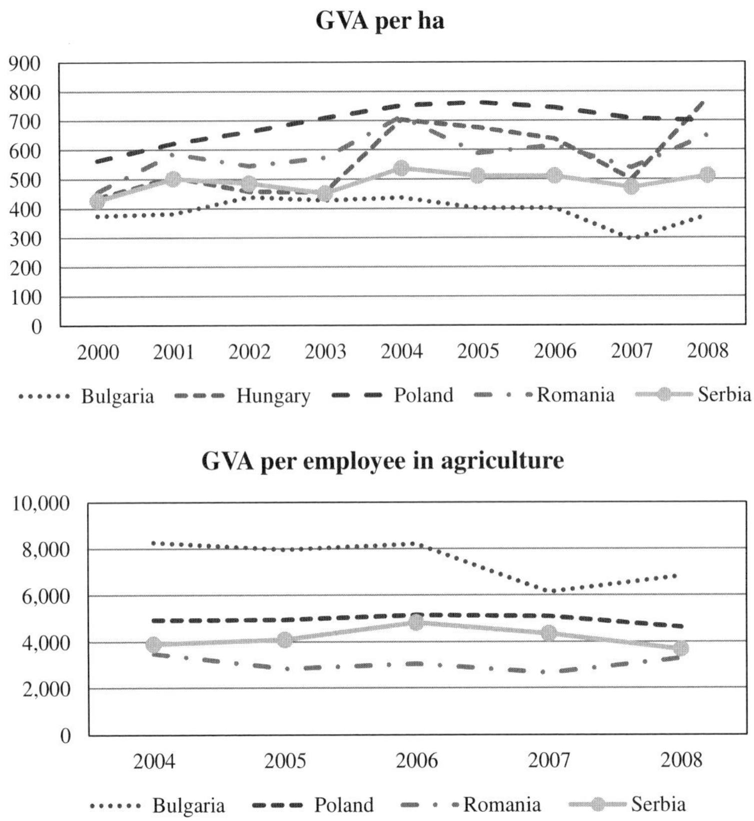

Figure 4.17. Gross value added (GVA) per hectare and per employee (in euro) in Serbia and several EU member states from Eastern Europe. Source: United Nations Statistics Division.

Developments in EU12 may be illustrative for developments in Serbian agriculture: establishing stable terms of trade, activating the land market, better access to foreign markets and stable sources of financing agriculture were the basic preconditions for agricultural productivity growth in all EU12. Reform of other parts of the economic system allowed the growth of employment in secondary and tertiary sectors, providing the opportunity for dynamic growth in labour productivity in agriculture. During the second half of the 1990s, all the candidate countries (except Romania and Bulgaria), recorded a decrease in employment in agriculture, both because of internal reform of their economic system and due to liberalization of labour markets and external migration of the rural labour force (Bogdanov *et al.*, 2010).

All EU-12 countries recorded a growth of GVA per hectare during the pre-accession period and in the first years of membership (see also Figure 4.17 for Hungary and Poland). More dynamic growth rates were related to the countries in which the initial value was low, which is primarily related to the Baltic countries. The growth of production in Romania and Bulgaria came later, which is attributed to the slow structural reform of agriculture and the rest of the economy, and slow withdrawal of funds for pre-accession assistance. Dynamics of GVA growth rates in the EU12 remains lower than the average EU15 mainly because of the negative trends that were at that time recorded by Romania and Bulgaria. More favourable trends than EU15 are in the Baltic countries and Hungary, while the growth rates achieved in Czech Republic and Slovenia are the same as of the old member states (Swinnen *et al.*, 2009).

4.8 Concluding remarks

The importance of agriculture for the Serbian national economy is high, which is confirmed by data on sector's share in GDP, employment and exports. This state is the result of a slow restructuring of the others sectors, but on the other hand it is the consequence of the rich natural resources available for agricultural production.

Available agricultural land per capita and per person employed in agriculture indicate significant benefits for Serbia in relation to other countries in the region.

The most intensive farming is practised in the northern part of country, which is characterized by flat plains and very fertile soils. The hilly and mountainous parts of Serbia are rich in pastures and meadows and have favourable conditions for fruit and wine production. Serbian agriculture is dominated by small and semi-subsistence farms with little produce to market. Some recent research estimate that about 60 per cent of farm holdings in Serbia do not have surplus production to sell or do sell on the market only infrequently. These farms generate low yields and lack specialization. Social reasons are an important explanation why these farms survive, as even though these farms are not economical yet for more than half of these households they do provide food security. In the northern part of the country there are large farms, well equipped, whose production is organized on the principles of modern management. However, regional differences in the structure of land, farm size, yields and production, are large and arise from heterogeneous natural conditions.

In terms of the value of agricultural production the most significant part is crop production, within which grains dominate. Livestock production is steadily declining and reached its lowest level in decades. In addition to the unfavourable economic position of this sector and low level of domestic consumption, the lack of consistent government support contributed to the decline of production. Frequent changes of support measures and the absence of a stable support system contributed to the creation unfavourable environment for the producers.

Yields and production are highly dependent on weather conditions, vary considerably and are still lower than the yield of the pre-transition period. Compared with other countries, Serbia has a higher yield of oilseeds and berry fruits, which are at the EU level. All other yields are close to regional averages and significantly below the EU27. The cause of low yields is insufficient use of inputs and lack of modern technologies and standards.

Prices of agricultural products in recent years have declined relative to inputs indicating a reduction in profitability in the sector. Prices for agricultural products are much lower in Serbia than the regional countries and EU27. Low prices of agricultural products in Serbia are the consequence of low labour costs and cheap land prices. This situation has led to a significant price-competitiveness of the Serbian agriculture compared to other countries in the region. However, because of low state support and low product prices, farmers' income is a low, contributing to the slow restructuring and modernization of the sector.

It is difficult to obtain a clear picture of farmers' income, due to a lack of data. Using GVA as a proxy for income, it appears that agricultural income varies significantly from year to year. Gross value added per person employed and per ha is lower than in neighbouring EU countries and close to the levels of Bulgaria and Romania.

Literature and websites

Agri-Livestock Consultants (ALC), *Dairy sector study for the IPARD programme, Republic of Serbia.* Final Report, September. Framework contract EuropeAid/127054/SV/Multi., 2010a.

Bogdanov, N. and D. Bozic, 'Review of agriculture and agricultural policy in Serbia.' In: *Agriculture in the Western Balkan Countries,* T. Volk (ed.), IAMO, Halle, Studies on the agricultural and food sector in Central and Eastern Europe, Vol. 57 (2010).

Bogdanov, N., P. Muncan and D. Bozic, *Family farms in Serbia - whether agriculture is an option to escape from poverty?(in Serbian)* 21st Scientific-expert conference of agriculture and food industry, Faculty of Agriculture and Food Sciences University of Sarajevo, Sarajevo. 2010.

Bozic, D., P. Muncan and M. Sevarlic, *An Assessment of the Competitiveness of the Dairy Food Chain in Serbia.* European FP7 Project AgryPolicy 'Enlargement Network for Agripolicy Analysis', 2009. Available at: www.euroqualityfiles.net/AgriPolicy/Report2.1/SerbiaAgripolicy D2-1.pdf

CAS (Cardno Agrisystems Consortium), *Dairy sector study for the IPARD programme Republic of Serbia.* Project: 2009/222867/2. Framework Contract EuropeAid, 2010.

Cvejić, S., M. Babović, M. Petrović, N. Bogdanov and O. Pavlović, *Social exclusion in rural areas of Serbia*,(in Serbian), UNDP, Belgrade, 2010.

Economic Institute, *The sectoral study on agriculture and food industry in the AP Vojvodina (in Serbian).* Belgrade, 2010.

Gill, S., V. Bogdanovic, C. Longo and J. Whealan, *Meat Sector Study for the IPARD programme.* EuropeAid/127054/C/SER/multi., 2010.

GoRS (Government of the Republic of Serbia), *First National Report on Social Inclusion and Poverty Reduction in the Republic of Serbia.* Belgrade, 2010.

GoRS (Government of the Republic of Serbia), Answer to the European Commission's questionnaire. Chapter 11 Agriculture and Rural development. 2011.

SORS, *Census of Population, Households and Dwellings (various publications).* Belgrade, 2002.

SORS, *Statistical Yearbooks and Bulletin Agriculture. Various editions 2000-2010.* Belgrade, various years.

Swinnen, J., K. van Herck and L. Vranken, *Agricultural productivity in transition economies, Choices.* 2009.

Volk, T., M. Rednak and E. Erjavec, 'Western Balkan agriculture and agricultural policy - Cross-country overview and comparison.' In: *Agriculture in the Western Balkan Countries.* T. Volk (ed.), IAMO, Halle, Studies on the agricultural and food sector in Central and Eastern Europe, vol. 57, 2010.

World Bank, *Integrating environment into agriculture and forestry progress and prospects in Eastern Europe and Central Asia.* Washington DC., World Bank, 2007.

Chapter 5

5 Serbia's rural population and agricultural workforce

5.1 Introduction

Rural areas in Serbia[1] are highly diverse in economic, social and demographic terms, due to differences in their geo-morphological characteristics (mountainous, hilly, plain areas), composition of population, economic structures, environmental conditions, transport accessibility, and employment and income possibilities. This chapter elaborates on the rural population and the people working in agriculture in Serbia. The chapter looks at urban-rural diversity in terms of labour force participation, unemployment, and the education levels of those living in rural areas and working in agriculture. In addition, this chapter describes the characteristics of poverty and social exclusion among rural population.

5.2 Rural population and agricultural labour force

During the last two decades a number of social, economic and political changes affected population numbers and structures. The war destruction caused a significant refugee inflow into the Republic of Serbia, whereas the economic and political instability led to considerable migration abroad. The complex outcome of these trends are only one component of aggregate tendencies, which together with natural trends as the consequence of biological depopulation affected population numbers in the Republic of Serbia in 2002, resulting as well as in regional disparity.

Until the beginning of the 1990s, rural areas in Serbia were characterized by a strong out-migratory trend due to the growth and development of the other sectors of the economy. This pulled labour from the agricultural sector. During the severe conflicts and war in the 1990s, population movements in rural areas were quite diverse. Mainly the mountainous and less fertile areas continued losing population. However, there was also immigration into rural areas by refugees from the other former Yugoslavian republics, the urban unemployed and retirees.

According to the most recent estimates by the Statistical Office of the Republic of Serbia, the population of the Republic of Serbia totalled 7,306,677 on 1 January 2010. Since the beginning of the 21st century, the population of the Republic of Serbia has constantly decreased and in early 2010 it was 220 thousand less in comparison with January

[1] The division into urban and other settlements in Serbia, found in the Census from 1981 to 2002 was based on the municipal decisions whereby municipalities gave a settlement the status of a town. It is enough for a settlement to have a general urban plan to be given the status of an urban area by a decision of the Municipal Assembly. All settlements that have not been proclaimed towns are otherwise classified as other. Statistical criteria are here obviously not respected (Bogdanov, 2007).

2000. In relative terms, the natural growth rate per one thousand inhabitants was -4.6. This decrease exceeded the population decline in several EU member states which also faced a negative population growth rate[1] (GoRS, 2011).

Currently, about 43 per cent of the population of the Republic of Serbia lives in rural areas. Since 2000, about 150,000 people or, on average, 2.7 per cent of the working-age populations (15-64 years) have left their previous place of residence each year. In total, during the period 1991-2002, population in rural areas of Serbia declined by 3.6 per cent compared to an overall 1 per cent decline in the country's total population.

Table 5.1. Settlements by type and populations number increase/decrease, 1991-2002.

	Total settlements		**Settlements with population decrease**		**Settlements with population increase**		**Settlements without population change**	
	all	**as % of total**	**all**	**as % of total**	**all**	**as % of total**	**all**	**as % of total**
				Serbia				
Total	4,706	100	3,847	81.8	834	17.7	25	0.5
Urban	169	100	57	33.7	112	66.3	-	-
Rural	4,537	100	3,790	83.5	722	15.9	25	0.6
				Central Serbia				
Total	4,239	100	3,566	84.1	649	15.3	24	0.6
Urban	117	100	36	30.8	81	69.2	-	-
Rural	4,122	100	3,530	85.6	568	13.8	24	0.6
				Vojvodina				
Total	467	100	281	60.2	185	39.6	1	0.2
Urban	52	100	21	40.4	31	59.6	-	-
Rural	415	100	260	62.7	154	37.1	1	0.2

Source: GoRS (2011).

In an analysis of population development at the municipality level, it appears that most of the urban municipalities show a population increase and most of the rural municipalities a population decline in the period 1991-2002 (Table 5.1). Municipalities with a positive population trend are mostly situated in the territory of Vojvodina and almost wholly in the South Backa and Srem Districts, as well as in the territory of the City of Belgrade. Figure 5.1 covers an even longer period showing the decrease of the population numbers over the period 1981 to 2011 showing the negative trend in the southern and eastern part of the country.

[1] In the same year, the natural growth rate in the European Union was 1‰, while the highest negative natural growth rates were recorded in Bulgaria and Latvia (both -3.6‰), Hungary (-3.4‰) and Germany (-2.3‰).

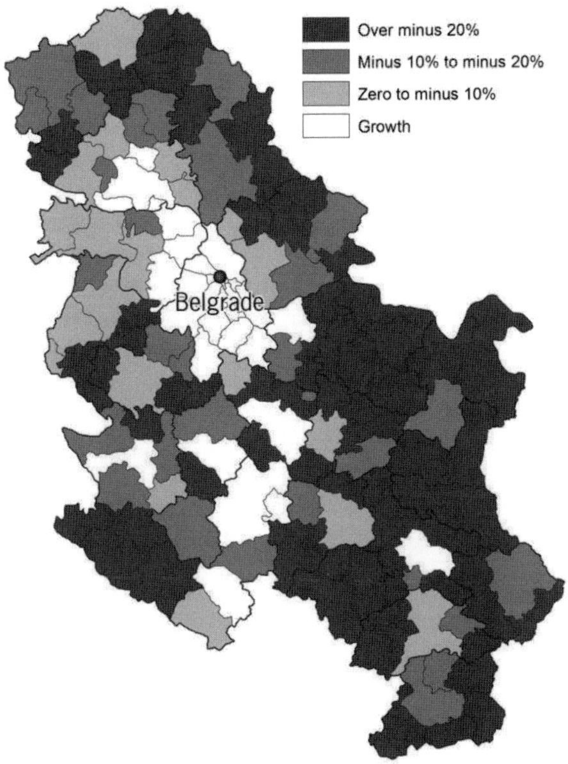

Figure 5.1. Decrease in the population number in Serbia (1981-2011).
Source: SORS, Census 2011.

The aging index, calculated as the number of persons aged 65 or older per hundred persons under the age of 15, is much higher in rural Serbia (1.35) than in urban Serbia (0.95). The total dependency ratio is also higher in rural areas: the proportion of dependent population groups (children 0-14 years of age plus persons aged 65 or older) compared to the working age population is 55 per cent, while that in urban areas is 43 per cent (SORS).

Unfavourable demographic trends are also determined by gender imbalances in rural areas. The share of the female population in the reproductive and working age (20-59 years) in the total rural population is 48 per cent, which is lower than the share of women in this age cluster in the urban population (52 per cent). The decline of the number of women in the rural population leads to falling birth rates, causes changes in the structure of agricultural production (reduction of production of vegetables, flowers and dairy products) and reduces the possibility of diversification of farm income through food processing.[1]

[1] The problem of imbalances of male and female population is also present in other European countries, but some differences appears in Northern Europe there is a problem of exclusion of the category of older women in single-

Unfavourable demographic trends have resulted in a negative education structure of the rural labour force, whose performance cannot meet the needs of the labour market. The share of illiterate people in rural areas according to the Census 2002 was 5.9 per cent (1.6 per cent in urban areas), and the share of people without primary schooling was 35.6 per cent (11.2 per cent in urban areas). These differences were more pronounced in Central Serbia (37.9 per cent versus 10 per cent) than in Vojvodina (29.5 per cent versus 14.2 per cent).

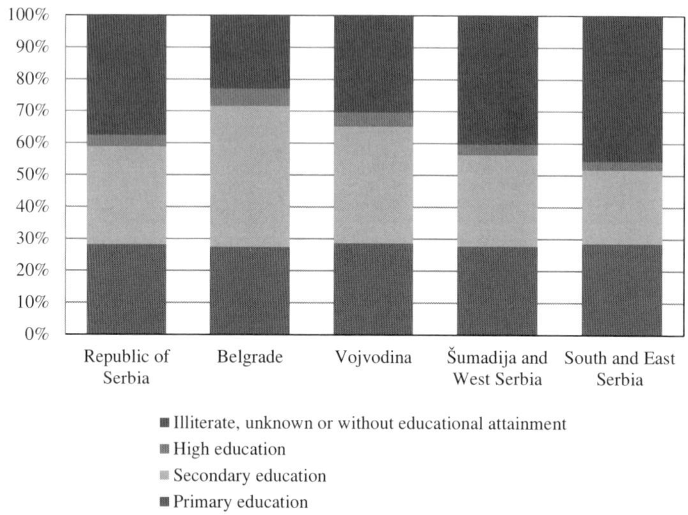

Figure 5.2. Rural population, 15+ by education.
Source: Authors' calculation based on Census (2002).

Modest knowledge and lack of additional skills characterizes the rural population in general, confirmed by LSMS 2007 data, according to which 97 per cent of rural population did not attend additional training and education courses, and 54 per cent have no special knowledge and skills (SORS, 2008). As a result of this poor education and training levels, knowledge and skills of the rural labour force are not appropriate for the requirements of the modern technology and adversely affect the overall capacity of the labour force in rural areas. This low quality of the workforce can be considered as one of the factors hampering economic development in rural areas because investors tend to avoid places without a high-quality and skilled labour force. On the other hand, this pushes outmigration of the higher educated population, as it is difficult to retain these people in rural areas without an attractive economic environment and jobs appropriate to their specific skills and preferences.

person households, while in Eastern Europe migration of young women because of barriers in the labour market have greater significance. [1] As many as 76 per cent of unemployed women have never benefited from any active labour market measures, while only 13 per cent have participated in a National Employment Service programme and stated that these programmes have not improved their employment prospects (Cvejić *et al.*, 2010).

The poor education structure, the low level of additional knowledge and skills and an insufficient coverage of active employment measures by the National Employment Service hinder the labour market opportunities of the 'rural' population and its competitiveness, in particular with regard to women and youth. Knowledge and new technology transfer in the area of food production takes place as part of the activities of the agricultural extension services, Rural Development Support Network, private advisers, trade companies and agricultural pharmacies. Other continuing education programmes are rarely accessible to the rural population.

5.3 Rural economy and rural labour market

Rural regions in Serbia cover 85 per cent of the total territory, with 55 per cent of the population living in that territory generating about 40 per cent of the country's gross domestic product (GDP). The industrial structure of Serbia's rural areas is highly dependent on the primary sector, especially on agriculture, and is still based on the depletion of natural resources. A high share of agriculture, the food industry and mining and a low participation of the tertiary sector in the GDP, as well as the low levels of entrepreneurship, are the main characteristics of the economic structure in Serbia's rural areas. They also reflect the underdeveloped structure of the rural economy.

The structure of rural household income indicates that income from agriculture has a relatively low share in total revenues, particularly in relation to salaries and pensions. Therefore, differences in living standards of rural household members are directly related to the possibility of employment outside the farm, which indirectly indicates that the performance of human capital and workforce, and their competitiveness in the labour market, have the greatest impact on the economic status of family farms (Figure 5.3). Regarding the regions, income from agriculture is more significant for households in Vojvodina. This is understandable, given that the average size of farms in this region and the number of market producers are significantly higher than in other parts of Serbia. On the other hand, the Region of Central Serbia is characterized by a greater share of income from pensions and income in kind, which are indications of a low level of development of the overall economy.

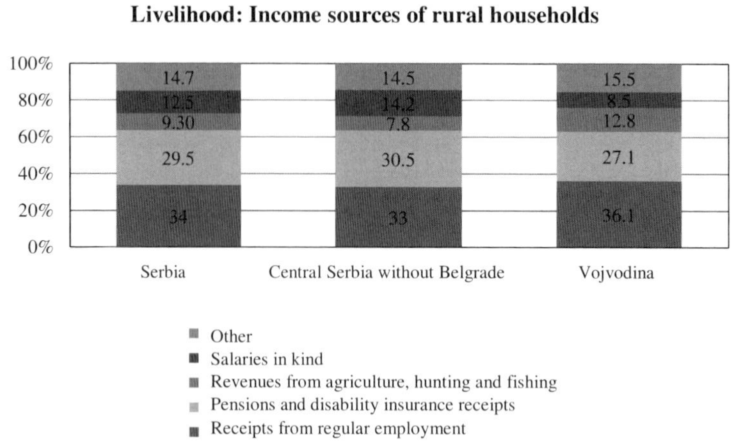

Figure 5.3. Livelihood: Income sources of rural households

Activity rates (49 per cent) and employment rates (40 per cent) of the rural population exceed those of the urban population (46 per cent and 36 per cent, respectively) (SORS Labour Force Survey 2010). Unemployment rates (16 per cent) and inactivity rates (52 per cent) of the rural population are below those of the urban population (21 per cent and 55 per cent, respectively). However, other performance indicators of the rural labour market, especially the share of vulnerable employment and the professional status of employees and others, are significantly worse among the rural population. According to the Labour Force Survey (LFS), in 2008 the share of vulnerable employment in total employment is as high as 48 per cent in rural areas, against 18 per cent in urban areas. The higher percentages of vulnerable employment among the rural population are due to the high percentage of farmers and unpaid family members, and a smaller share of waged employees in the rural compared to the urban population. The LFS 2008 data suggest that almost 40 per cent of all employees in rural areas had the status of farmer or unpaid household labour (SORS, 2008). On the other hand, the proportion of the waged employed in total rural employment (52 per cent) was below that in urban employment (82 per cent).

Table 5.2. Labour market status and vulnerable employment (% of total labour force).

	Total	Urban	Rural
Waged employees	68.8	81.7	52.4
The owners of enterprises and shops	7.7	9.9	4.9
Self-employees	4.1	4.6	3.5
Agricultural workers	11.2	2.3	22.6
Unpaid family workers	8.2	1.6	16.7
% of vulnerable employment	31.2	18.4	47.7

Source: SORS, LFS (2008).

The most significant manifestation of changes in the rural labour market in Serbia are the high unemployment rates, especially of the younger population, and the high rates of permanent migration of young women. The economic crisis has highlighted the figure of unfavourable trends in the rural labour market – the most prominent change being the decline in the employment rate of rural youth and the highly educated population.

Table 5.3. Rural and urban population activity rates in Serbia, broken down by some demographic indicators of the population aged 15-64, 2008 and 2009 (%).

		LFS 2008		**LFS 2009**		**2009/2008, percentage points**	
		Urban	**Rural**	**Urban**	**Rural**	**Urban**	**Rural**
Total (15-64 years of age)		60.4	66.4	59.5	62.6	-0.9	-3.8
Sex	Males	67.9	75.5	66.1	73.1	-1.8	-2.4
	Females	53.5	56.9	53.4	52.0	-0.1	-4.9
Education	Lower	55.9	29.7	29.6	51.3	-0.1	-4.6
	Secondary	63.7	74.0	63.1	70.3	-0.6	-3.7
	Higher	82.6	83.8	80.0	77.7	-2.6	-6.1
Age structure	15-24	26.6	38.2	25.3	32.7	-1.3	-5.5
	25-49	82.9	82.8	82.6	79.1	-0.3	-3.7
	50-64	47.2	58.6	45.0	56.6	-2.2	-2.0

Source: SORS, LFS (2010), own adaptation.

The absence of active employment initiatives by the National Employment Service reduces the chances of the rural population in the labour market. On the other hand, the rural population with low levels of education can participate in public works programmes.[1]

Nearly half of the rural labour force is employed in the agricultural sector and about one third in the services sector (Table 5.4). In urban areas, the services sector is the largest employer, involving about two thirds of the urban labour force. In rural areas, the structure of employment by sector of activity is dominated by agricultural workers, while the share of employees in the service sector is well below the average for the Serbian economy.

Table 5.4. Employment by sector in rural and urban areas in Serbia, 2010 (% of total labour force).

	Total	Urban	Rural
Agriculture, forestry and fishing	21.9	4.1	43.6
Mining and quarrying	1.1	1.1	1.1
Manufacturing	17.1	18.6	15.3
Electricity, gas and water supply	1.4	1.8	0.9
Water supply and waste management	1.6	1.9	1.3
Construction	5.1	5.5	4.7
Wholesale and retail trade, repair of motor vehicles	13.5	15.9	10.7
Transport, storage and communication	5.4	6.4	4.2
Hotels and restaurants	3.2	4.0	2.2
Information and communication	1.9	2.8	0.9
Financial services, insurance	1.9	2.8	0.9
Real estate, renting activities	0.2	0.2	0.0
Scientific, educational and technical work	2.3	3.8	0.5
Public administration and social services	2.3	3.8	0.5
Other community work, social security	4.8	6.3	3.1
Education	7.1	9.7	4.0
Health and social work	6.2	8.1	4.0
Art, fun recreation	1.3	2.0	0.5
Other services	1.7	2.1	1.3

Source: SORS, LFS (2010).

[1] As many as 76 per cent of unemployed women have never benefited from any active labour market measures, while only 13 per cent have participated in a National Employment Service programme and stated that these programmes have not improved their employment prospects (Cvejić *et al.*, 2010).

5.4 Employment in agriculture

As Table 5.5 shows, in 2002 there were some 454,000 active farmers in Serbia, which equals about 13 per cent of the total active population of Serbia (which was some 3.4 million in 2002). The total number of active farmers decreased by 45 per cent in the period 1991-2002. Over the same period the relative importance of off-farm activities increased (from 23 per cent in 1991 to 28 per cent in 2002). In absolute terms, however, the number of people engaged in off-farm activities also decreased over this period. During the same period the number of farm households who received a personal (non-farm) income more than doubled. The largest 'agrarian exodus' took place in Central Serbia where the number of active farmers decreased by 47 per cent against 24 per cent in Vojvodina.

Table 5.5. Farm households members, by activity.

	Total number of farm members	Active farmers		Off -farm Active members		Members with personal income		Supported members	
		No	% of total	No	% of total	No	% of total	No	% of total
				Serbia					
1991	3,557,867	816,200	22.9	799,926	22.5	354,484	10	1,289,369	36.2
2002	2,536,378	454,732	17.9	705,070	27.8	526,232	20.7	845,238	33.3
Index 1991 = 100	71.3	55.7		88.1		148.5		65.6	
				Central Serbia					
1991	2,586,118	689,630	26.7	552,624	21.4	234,486	9.1	880,844	34.1
2002	1,915,757	365,914	19.1	505,455	26.4	411,198	21.5	628,848	32.8
Index 1991 = 100	74.1	53.1		91.5		175.4		71.4	
				AP Vojvodina					
1991	803,392	116,763	14.5	206,508	25.7	107,662	13.4	318,355	39.6
2002	620,621	88,818	14.3	199,615	32.2	115,034	18.5	216,390	34.9
Index 1991 = 100	77.2	76.1		96.7		106.8		68	

Source: Bozic and Bogdanov (2005). Note: The differences from 100% are farm-holding members working abroad for the period of maximum one year. In 1991 there were 8.4% working abroad and in 2002 only 0.3%.

Box 5.1 Definition of categories of agricultural population.

Individual farmers are part of the active agricultural population and perform their profession on their own agricultural holding or on the family's agricultural holding.

Active members outside their own farm are all active persons who perform a profession outside the agricultural sector. This group involves also the active persons who are farmers by profession but are employed outside their own or family farm.

Members with personal income are persons who are supported independently by regular income such as pension, property income, social welfare and other regular income.

Supported members are persons without their own income, as well as persons with insufficient income and they are supported by their parents or other persons or legal entities. This includes supported persons whose supporter was abroad during the time this registration was made.

Members temporarily employed abroad are citizens of Serbia who during the time of registration were temporarily employed abroad by a foreign employer or independently.

Source: Census (2002), SORS.

When comparing the data for the two last Population Censuses (1991-2002) it is clear that Serbia has experienced a dramatic reduction in the total farming population in that decade. This 'agrarian exodus' resulted in significant demographic changes in the structure of farms and the composition of family farm members, which are reflected primarily in the age structure of the farming population. According to the records from the last Census, the agricultural population has a higher percentage of people older than 45 years than in the total population (see Figure 5.4). The proportion of young farmers (up to an age of 19), as well as those in the most productive age (20 to 49) is decreasing. The educational structure of the active agricultural population of Serbia is very unfavourable. Many farmers have not completed primary education, while the proportion of illiterate persons exceeds that of persons with secondary and higher education. Reproduction of labour on family farms is seriously challenged. The results of the Census 2002 show that about 120 thousand households had only one member.

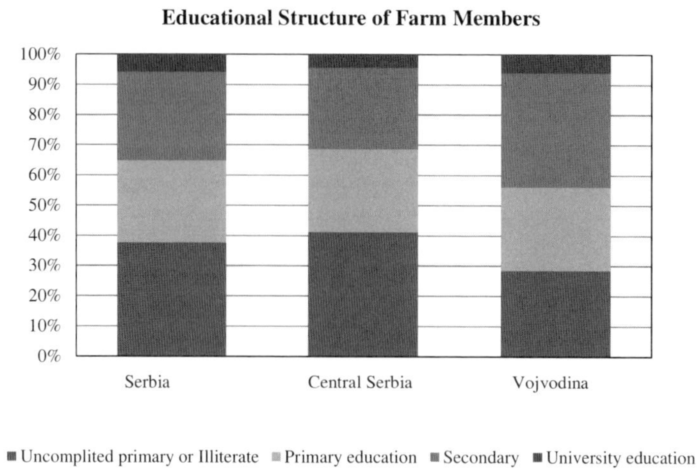

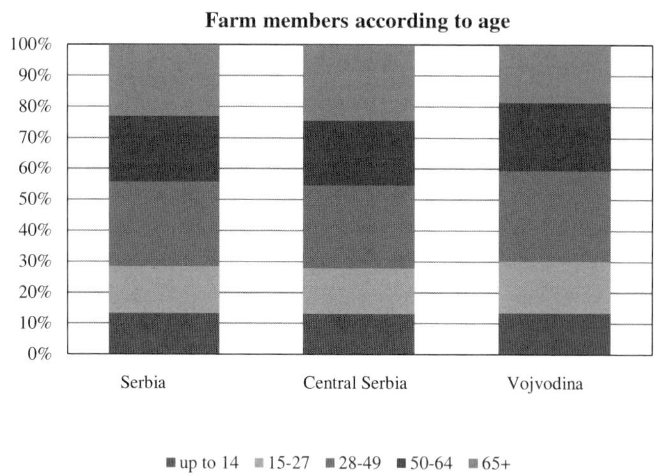

Figure 5.4. Age and educational structure of farm members in Serbian regions. Source: Authors' calculation based on Census (2002).

5.5 Rural poverty

Poverty in Serbia is strongly dependent on the characteristics and intensity of rural poverty. Interpretation of rural poverty in Serbia as well as poverty of agricultural households is hampered significantly by the lack of precise statistical definitions and classifications. Namely, a definition of the rural area in Serbia based on statistical indicators does not exist, while the definition of an agricultural household may vary according to the agricultural resources used. Moreover, agricultural households are not *a priori* located in rural areas.[1] Types of agricultural households are defined according to income sources of the family members.[2] Apart from lack of precise statistical definitions, there is also a lack of other typologies of rural households and agricultural households; this hinders thorough analyses of poverty parameters of different social and economic groups of households, in general, and of agricultural households in rural areas in Serbia, in particular.

The most recent research into poverty and living standards indicates that poverty in Serbia is:

more widespread in rural areas than in urban areas (9.6 per cent as opposed to 4.9 per cent);
strongly linked to the performance of the agricultural labour force and the rural labour market, as one half of all the persons employed in agriculture live in 'poor' households;
more prevalent in southern and hilly/mountainous regions in comparison to northern plain regions; Vojvodina is the region with the lowest percentage of population encountering at least one dimension of poverty; Western Serbia has the lowest percentage of financially poor population (33 per cent), and South-Eastern Serbia the highest (41 per cent) (Bogdanov and Petrovic, 2010).

Owing to high initial values of rural poverty rates, the decrease of poverty until 2008 was more pronounced in rural areas than in urban areas (Table 5.6).

Table 5.6. Poverty indicators by settlement types in Serbia, 2006-2009.

	2006	2007	2008	2009	Index 2009/2008
Absolute poverty line (SRI)					
% poor	8.8	8.3	6.1	6.9	113.1
% poor by type of settlement					
Urban area	9.5	10.1	5.0	4.9	98.0
Rural area	20.6	17.6	7.5	9.6	128.0
Relative poverty line					
% poor	14.4	13.4	13.2	13.6	103.0
% poor by type of settlement					
Urban area	9.5	10.1	10.9	9.1	83.5
Rural area	20.6	17.6	16.1	19.5	121.1

Source: Government of the Republic of Serbia Social Inclusion and Poverty Reduction, Unit and the Statistical Office of the Republic of Serbia (2010).

The lowest rural poverty rate was recorded in 2008, which can be attributed to the increased agricultural commodity prices. However, rural areas of Serbia responded to the

[1] An agricultural household owns and works on at least 10 acres of land, or with a plot of land smaller than 100 m^2 if they possess: at least one cow and a calf, or one cow and a heifer, or one cow and two adult animals of the same species, or five sheep, or three pigs, or altogether four adult heads of sheep and pigs, or 50 chickens, or 20 beehives.

[2] Agricultural holdings are those where all income is derived from individual agricultural workers on the holding. Non-agricultural holdings are those where income is derived from a member, or members of the household who do not work in agriculture; or they do, but outside their own or family holding; or income comes from a pension, other possessions, social welfare or some other source of regular income. Mixed farm holdings are those where income comes both from agricultural and from non-farm sources. Farm holdings without income are those where the income source is unknown or it comes from supporters (including legal entities) that do not belong to the household.

economic crisis promptly and strongly: the growth of overall poverty recorded in 2009 was for the most part a result of dynamic growth of rural poverty itself. In comparison to 2008, the percentage of the poor in rural areas increased by 28 per cent in 2009, whereas in urban areas their number remained virtually the same. Deterioration in the farmers' position and growth of unemployment were the most striking manifestations of the global crisis impact on rural areas. The growth of rural poverty will contribute to a slowdown of structural reforms in agriculture and lead to a deepening of the already big *differences* in *development* between *rural and urban* areas of Serbia. Regional differences in rural poverty rates are significant and follow the relationship that exists between regions in terms of overall poverty: the most unfavourable situation being in southern and eastern Serbia, and the more favourable being in Vojvodina (Bogdanov and Petrovic, 2010).

A recent survey of social inclusion of rural population shows that agricultural households have the highest percentage of financial poverty (Cvejic *et al.*, 2010). According to this study, poverty is significantly related to farm size; in Serbia and all the regions analysed, except South Eastern Serbia, a higher poverty rate was observed among large farm households whose production costs in 2009 were high (see Table 5.7). According to the type of household, farms with mixed sources of income are in a much better position than the agricultural households whose members have no other source of income.[1]

Table 5.7. Percentage of family farms below poverty line.

	Serbia	**Vojvodina**	**Central Serbia**	**West Serbia**	**South Eastern Serbia**
Total	38.7	39.36	38.73	33.22	42.9
According farm type					
Non agricultural	33.67	33.83	23.27	29.07	42.92
Part-time farms	30.54	38.89	37.9	18.1	29.41
Agricultural	53.44	47.37	61.61	51.92	51.22
According farm size					
Small (up to 1 ha)	40	38.55	31.88	32.2	52.81
Medium (1,01-5,00 ha)	34.62	29.59	37.97	28.38	39.39
Large (above 5,01ha)	43.76	49.5	43.17	42.05	40.21

Source: Cvejic *et al.* (2010).

Other visible manifestations of poverty and exclusion of the rural population in Serbia, which were found in the survey, are as follows (Cvejic *et al.*, 2010):

- The problem of early school leavers in rural areas is very distinct, considering that 44 per cent of respondents aged 15 and over had left school before graduating from high school. Elementary schools exist in most villages, but the kindergartens, cultural centres and libraries are poorly represented, except in Vojvodina. The availability of additional, alternative educational programmes is weak.

[1] The methodology of poverty measurement in this study is set in accordance with EU methodology, which means that poverty is measured by income rather than consumption. The sample was prepared in accordance with OECD standards for defining rural areas. Therefore the image of rural Serbia in terms of social exclusion and poverty is different from the common one, based on official statistics, which does clearly separate rural settlements.

- The coverage of rural population by social insurance (pensions and health) is not sufficient. From the group of elderly households who do not have a right to a pension, twice as many are exposed to poverty than elderly households receiving a pension. Over half of the respondents of working age do not contribute to pension funds, they remain completely unprotected in the case of injury and disability, and they are exposed to extremely high risks of poverty in old age.
- As many as 13 per cent of respondents have no health insurance and face serious difficulties in accessing adequate health care services. Employed persons in agriculture and the unemployed are more often not covered by health insurance than employees in other sectors.
- Access of the rural population to social protection services is very weak. Despite the high percentage of poor, only 1.9 per cent of households are beneficiaries of child allowance and individual/family allowance. Older residents have the hardest position, as they face serious obstacles in the provision of social services for elderly and disabled persons.
- Half of the unemployed used to be employed outside the farm, however they have little chance of re-employment. This is indicated by a high share (over half) of the long-term unemployed.
- The unemployed in rural areas are not only excluded from the labour market, but also from the measures that aim to improve their employability. These measures have been implemented in various forms through the National Employment Service. Unfortunately, these are hardly implemented in rural areas. 88 per cent of the unemployed within the sample had not benefitted from any kind of support provided by NES, and only 2.7 per cent participated in some of the programmes. In addition, only 3.4 per cent of the unemployed have attended additional educational programmes. Those who did not attend prequalification programmes cited a lack of money as the most common reason (43 per cent), followed by lack of interest (28 per cent), remoteness and lack of adequate offers of courses (9 per cent each).

5.6 Rural infrastructure

Mapping rural infrastructure and making an analysis of the needs for various types of services in Serbian rural areas have not been carried out to date. According to some findings (Bogdanov, 2007) the existing situation concerning the coverage of villages by basic utility systems (services) is such that over 80 per cent of villages have access to the basic services: road network, electricity, and telephone network (fixed and mobile). Water supply systems, gas line systems, sewage systems and waste dump sites (landfills) constitute systems that are much less present in villages in Serbia. In general, utility service infrastructure in rural areas in Serbia is highly underdeveloped and neglected, and thus the provision of services does not meet the needs of the rural population. Rural inhabitants themselves put the improvement of the utility service infrastructure, and especially the water supply systems, sewage systems, electricity and road networks as priority issues, even when compared to their economic problems.

According to the LSMS (SORS, 2007), the water supply problems in rural areas vary according to the geomorphology, the village location and the available water supply alternatives. Villages that are closer to urban areas are eventually connected to their water supply lines, if they manage to organize themselves and pay for the connection. Data indicate that villages that do not have access to a central water supply rely on wells as their primary source of water.

The First National Draft on Social Inclusion and Poverty reduction in RS (2010) states that:

The school network is not in line with demographic changes, which indicates that there is still a high number of remote classrooms with a smaller than average number of pupils in the areas which are hard to reach, and which are without basic equipment and with a low quality of education.

There is a big gap in access to IT technologies, which is the consequence of inadequately developed infrastructure. The difference can be seen when the household presence of computers in rural areas (38.9 per cent) is compared with their presence in urban areas. Compared with 2009 this gap is slightly reduced. In urban areas the pace of annual growth of the number of computers is 2.7 per cent, while this is 4.7 per cent in rural areas.

Rural infrastructure demands significant investments in new systems, in their revitalization and expansion. It is an important factor to improve the quality of life of the rural population and it also has a stimulating effect on potential investors. The municipal governments have the greater part of administrative responsibility for infrastructure investments, yet the financial resources needed are insufficient for the scope of investments that need to be made.

Literature

Bogdanov, N., *Small rural households in Serbia and rural non-farm economy.* Belgrade, UNDP, 2007.

Bogdanov, N., *Agriculture. living standards measurements study-Serbia 2002-2007 (LSMS).* Statistical Office of the Republic of Serbia, Belgrade, 2008.

Bogdanov, N. and M. Petrovic, *Financial poverty and material deprivation in rural Serbia. Poverty and Social Inclusion in the Western Balkans.* Brussels, 2010.

Božić, D. and N. Bogdanov, 'Changes in socio/demographic characteristics of farm households members in Serbia.' In: *Family farms in Serbia during transition. Dj. Stevanović and D. Živković.* Belgrade, Institut za agroekonomiju, Poljoprivredni fakultet, pp. 68-90, 2005.

Cvejić, S., M. Babović, M. Petrović, N. Bogdanov and O. Pavlović, *Social exclusion in rural areas of Serbia (in Serbian).* Belgrade, UNDP, 2010.

GoRS (Government of the Republic of Serbia, 'Social inclusion and poverty reduction.' In: *Monitoring Social Inclusion in Serbia.* SORS, Belgrade, 2010.

GoRS (Government of the Republic of Serbia), *Answers to the European Commission's Questionnaire.* 2011.

SORS, *Census of population, households and dwellings (various publications).* SORS, Belgrade, 2002.

SORS, *Living Standard Measurement Survey - Serbia 2002 - 2007.* SORS, Belgrade, 2008.

SORS, *Living Standard Measurement Survey - Serbia 2010.* SORS, Belgrade, 2010

SORS, *Household budget survey 2008, 2009 and 2010.* SORS, Belgrade, various years.

SORS, *Survey on labour force. 2008, 2009 and 2010.* SORS, Belgrade, various years.

SORS, *Census2011.* SORS, Belgrade.

Chapter 6

6 The agri-food supply chain

6.1 Introduction

If Serbia enters the EU, the country will be part of the internal market. The impact of open borders on the Serbian food chain depends on its competitive position, which is affected by the chain's structure and performance. This chapter discusses the Serbian food chain, starting at the consumer level and moving up the chain all the way to the suppliers of farmers. The analysis presents key indicators of structure and performance to evaluate the agri-food supply chain's strengths, in a comparative analysis with benchmark countries having similar supply chain structures, such as Romania, Bulgaria, Hungary and Poland, and also making the comparison with the EU27 average. The sections in this chapter analyse in turn the food consumption patterns, food retailing and wholesaling, food processing, primary agriculture and, finally, input suppliers. Structural features of the Serbian agri-food chain have implications for institutional arrangements in the field of agricultural production and trade. This chapter concludes with a section on how these linkages between farmers, processors, traders and the retail sector can be characterized in Serbia.

6.2 Consumption patterns

The Serbian household consumption expenditure is USD 4,300 per capita in 2009 (see Figure 6.1). This level is almost the same as in the newest EU member states, Bulgaria and Romania, yet only half the level in Poland and Hungary. The EU27 average household consumption expenditure is USD 20,000 in 2009. The developments in expenditure in Serbia have been rather similar to those in Bulgaria and Romania, which is not surprising as these countries have more or less similar income per capita levels. The data on household final expenditure is derived from the national accounts and presents the expenditure incurred by residents' households on consumption goods and services.

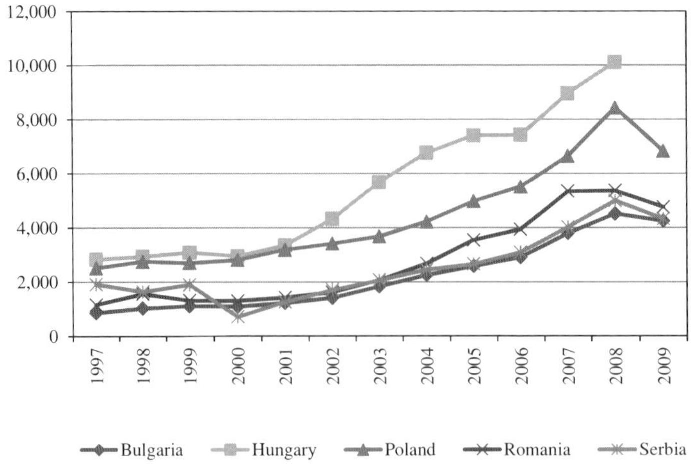

Figure 6.1. Household final consumption expenditure (USD/capita).
Source: World Bank.

Data on the share of food in the total household expenditures in the benchmark countries show that the higher the income the lower the share of food in the total average consumption expenditures: food and non-alcoholic beverages have a share of 15 per cent in the EU27 and 44 per cent in Romania (Eurostat, HBS). Data on Serbian food expenditures as a share of total expenditures reveal that the figures are close to those in Romania: 40-41 per cent in the years 2007-2010 (SORS, statistical yearbooks). This can also be estimated indirectly using the value of domestic demand for the food, beverages and tobacco industry, which is around €5 billion in 2007 to 2009 (GoRS, 2011). With around 7.3 million inhabitants in 2.5 million households, the annual expenditure would be between €1,800 and €2,000 per household, just below the level of Romania and Bulgaria.

The food expenditures in Serbia, calculated in this way, do not included informal food purchase, e.g. directly from farmers or other informal channels such as the farmers' market, as these purchases are not registered. The informal market is, however, a significant part of the food chain. Gill *et al.* (2010: 118-121) estimate that one third of the beef, two thirds of the poultry, more than half of all pig meat and around 90 per cent of sheep meat production is consumed at home or 'sold through informal non-recorded trading channels'. Informal markets are important because people's limited income restrains them from buying food in shops, while in remote areas informal farmers' markets are the only alternative for buying food. Gill *et al.* (2010: 87) suggest that the contribution of own consumption to household income is between 30 and 60 per cent in those parts of Serbia where semi-subsistence farming prevails. The figures of Gill *et al.* refer only to meat consumption. However, given the fact that data for the entire country indicate that in Serbia only about 40 per cent of family farms produce market surpluses (Bogdanov, 2007; LSMS, 2007; Bogdanov and Petrovic, 2010) the estimate of Gill *et al.* for the percentage of own consumption in the household income may hold for the whole primary agricultural sector.

Food expenditures in the four benchmark countries are largely on meat, cereal based products, dairy and vegetables, which account for 70 to 80 per cent of the total expenditures on food (Eurostat HBS). In order to compare food consumption patterns in Serbia with those in the benchmark countries, we have to rely on FAO data on consumption quantities.[1] The diet in Serbia differs slightly from that in Romania and Bulgaria: Serbians consume much less milk, cereals and vegetables, but much more fruit. Compared to EU27 levels, Serbia has a relatively low consumption of milk (products) and cereals, yet a high consumption of fruits and meats (measured in primary equivalents) according to these FAO data (Figure 6.2).

6.3 Retailing and wholesale of food

6.3.1 Benchmark countries

Detailed information on the structure and performance of retailing for EU-countries is available in the Eurostat SBS database, such as where people buy their food products and how much, and retail margins. Figure 6.3 shows that the number of specialized shops such as butchers, greengroceries or bakeries is at or below 30 per cent of the total number of retail shops, and that their share in turnover and employment is below 20 per cent, respectively, in all four selected EU countries. This indicates that the majority of food is bought in supermarkets. The gross operating margin is between 2 per cent (Hungary) and 7 per cent (Poland) of turnover.[2] The difference between the specialized and non-specialized food shops is much smaller than the differences between countries for the same retail outlet. Roughly 80 per cent of the enterprises have a share of 25 per cent in the turnover, or in other terms the 20 per cent largest enterprises have 75 per cent of the total turnover. For the EU27, 10 per cent of the largest enterprises have 90 per cent of the turnover (Eurostat, SBS data). These data show that the retail market structures are highly concentrated in the whole EU, and also in those countries which entered the EU only recently and are in the lower range of GDP per capita among the member states.

---

[1] Food consumption data from the Serbian Household Budget Survey, published by the SORS, show rather different levels of consumption (in kg) per capita from what is shown in the FAO data: cereal consumption is much higher and meat consumption much lower than FAO data suggests. The reasons for these differences may be found in the methodology and in the sample of the survey.

[2] This gross operating margin is the turnover minus costs of inputs, labour and credits. This means that this margin is the rent for own capital and labour.

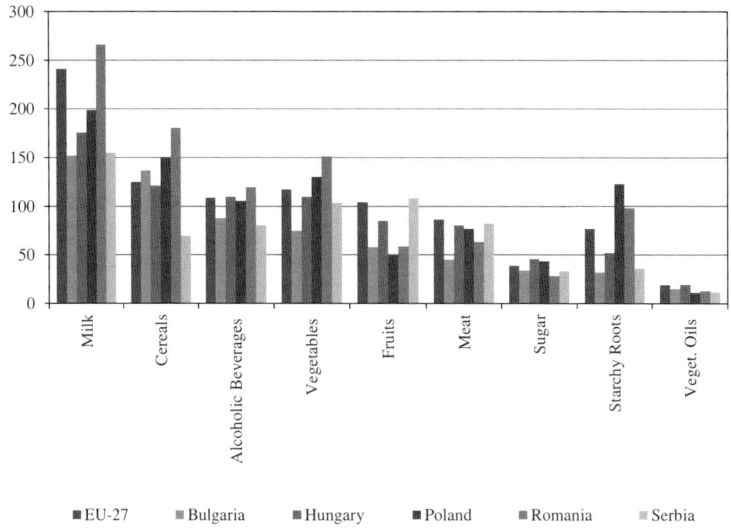

Figure 6.2. Consumption in kg/head/year in primary equivalents in 2007. Source: FAO, FAOStat.

6.3.2 Evidence from Serbia

Food retail in Serbia is still fragmented. Most recent data available indicate that small grocery stores are the most popular format, accounting for some two-thirds of sales on the formal food market. There are around 30,000 independent stores operating around the country (Bord Bia, 2008). In the bigger cities, though, the supermarket format is developing. The emergence of supermarket chains is much supported by the inflow of capital from foreign supermarket chains since the beginning of the 2000s. Currently, foreign owned supermarket chains are the largest in the country (see Table 6.1). Recently, the Delhaize Group, a Belgium-based grocery and supermarket chain, has acquired Delta Maxi Group, by far the largest food retailer in the country (RBR, 2011). Mercator, Slovenia's largest food retailer has taken full control of the Serbian chain Rodic.

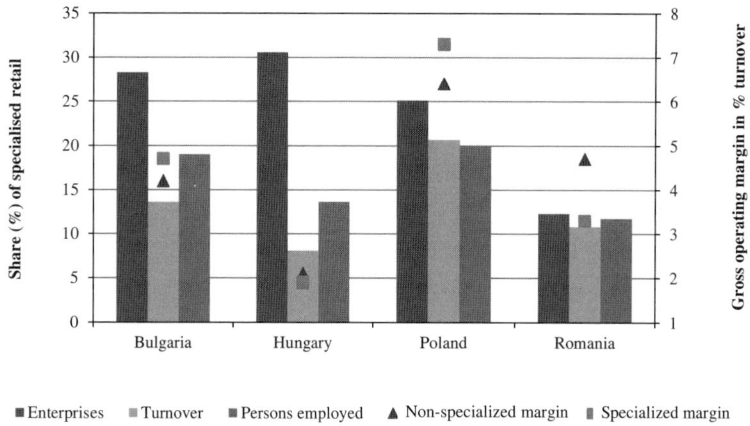

Figure 6.3. Shares of specialised retail in total retail of food and gross operating margin in per cent turnover in 2008. Source: Eurostat, SBS, Annual detailed enterprise statistics for trade (NACE Rev.2 G).

Table 6.1. Supermarket chains in Serbia.

Retailer	Country owner	Total sales (€m)	Grocery sales (€m)	Grocery retail market share (%)	No of stores
Delta Holding (Maxi) (1)	Serbia	1,720	1,238	16.8	325
Mercator/Rodic	Slovenia	342	320	4.3	59
Metro	Switz/Germ.	187	187	2.5	5
Idea (Konzum)	Croatia	164	164	2.2	41
Simpo (Si market) (2)	Serbia	15	156	2.1	171
Interex (Intermarche)	France	75	75	1.0	6
Super Vero (Veropoulos)	Greece	69	69	0.9	3

Note: (1) Delta Holding has been taken over by Delhaize, Belgium in 2011; (2) Simpo has become part of Delta Holding in 2010, and hence, is owned by Delhaize since 2011.
Source: Web search and IGD Analysis, Country Presentation Serbia, 2008 cited from Bord Bia (2008).

With the increasing share of supermarkets, the private label brands are also emerging. Private label brands are generally owned by, and sold through, a specific chain of stores (Interbrand, 2011). They are also known as store brands. Consumers purchase private brands because they appreciate the good price-quality relation. The prices can be up to 35 per cent lower compared to specific manufacturer brands without much quality differences. The advertisement costs of world brands are high to maintain the image of quality and exclusiveness. Private labels assist in building consumer loyalty towards a supermarket chain (Beneke, 2010). The market shares of private labels are rising, not only in the rather saturated food markets in Western Europe and the USA, but also in countries like Hungary and Russia, where supermarket chains rapidly develop and increase their share of the food retail market.

The growth of the turnover of private labels in Serbia is high and connected to the infant phase of the development to the so-called fast growing markets. International retail chains in Serbia exert significant influence on this development, introducing labels they already developed in other markets. Private labels availability for consumers is the highest in Belgrade and in the Vojvodina region, where over 60 per cent of their offer is concentrated in the 10 largest cities (Koncar *et al.*, 2010). Delta Tempo, Idea and Mercator Center have in their assortments over 1,000 private label products, while Interex and Metro sell about 300 to 500 private label products (Koncar *et al.*, 2010). Little information on the origin of the private labels is found. Koncar *et al.* reveal that Metro has 125 (food and non-food) articles from Serbia and 200 foreign brands. It would offer the Serbian agri-food supply chain great perspectives if it becomes a preferred supplier of these products. For this to happen it is necessary to establish close contacts with the supermarket chains and to comply with the standards they require.

Yet, next to the formal market for food, a significant part of food is acquired through informal channels directly from the farm and at farmers' markets, while on-farm use of own-produced products is also important. Available data suggest that overall one third to half of the agricultural production is consumed at the farm or sold on farmers' markets. Goss *et al.* (2010) found that one third of the dairy production is consumed at the farm or directly sold on the farmers' markets (see Table 6.2), while Gill *et al.* (2010) estimate that the number of pigs slaughtered off-farm is 30 per cent at the most, implying that 70 per cent of the pig meat is for household consumption or direct sales.

Table 6.2. Serbian marketing channels of milk and dairy products, sales in litres milk equivalent.

Market channel	Liquid milk		Dairy Products		Total	
	Million litre	Per cent	Million litre	Per cent	Million litre	Per cent
Household use	96	6.6	93	6.4	189	13.0
Direct sales	90	6.2	106	7.3	196	13.4
Green markets	-	-	113	7.8	113	7.8
Shops and supermarkets total	380	26.1	580	39.8	960	65.8
Of which Small shops	171	11.7	261	17.9	432	29.6
Total	566	38.8	892	61.2	1458	100.0

Source: Goss *et al.* (2010), pages 134/135.

Few data are available about Serbian wholesale food and agricultural commodities. The scarce information suggests that wholesale as a system of business-to-business trade is little developed. Yet in the Serbian food trade there are some big players such as the German-based Metro, Velpro (from Croatia) and Serbian DIS (Koncar *et al.*, 2010). These companies serve as an intermediary between suppliers and buyers, and therefore play a crucial role in the efficient operation of the market. With the opening of its sixth distribution centre in Subotica in October 2010, Metro announced that it will procure about 80 per cent of the food products from local suppliers and producers and therefore inject a positive stimulus in the domestic industry. Of course, local suppliers will only benefit from such opportunities if they are able to fully comply with the quality standards required. This, among other things, implies the adaptation of production units to the HACCP standard, something which is still to be applied within many food processing firms in Serbia.

6.4 Food processing

Generally speaking, there are not many studies on the Serbian food industry, and those which are available refer to difficulties in finding appropriate data. However, Chapter 20 (Enterprise and industrial policy) in the Answers to the Commission (GoRS, 2011) provides a rich variety of key data to picture the economic situation of the food industry in Serbia. This will be the main source of the analysis below. Not all data can be compared with our benchmark countries, but where possible and relevant figures will be presented to illustrate the situation in Serbia in a comparative analysis.

Importance of the food industry

Primary agriculture is relatively important for the Serbian economy, in terms of both GDP and employment (see chapter 2). Manufacturing contributes only 18 per cent to Serbia's total GDP, which is rather low compared to the benchmark countries in this chapter. However, within manufacturing, the food, beverage and tobacco industry adds a significant 21 per cent to the GDP of total manufacturing, pointing at a more important position within the overall industry than in Bulgaria, Hungary and Poland. Together, the agricultural and food processing industries account for 15 to 20 per cent of Serbia's GDP.

Structure of the food industry

The number of enterprises registered in the food, drinks and tobacco sector (NACE 15 and 16) is about 3,300 (GoRS, chapter 20, 2011: 448-449). The number of enterprises has increased since the mid-2000s. Around one third of this number can be found in the meat and bread manufacturing. The food sector employs almost 74,000 people (see Table 1 in Annex to this chapter). Most employment is also in the meat and bread branches, followed by the mineral waters and soft drinks industry, the dairy processing industry, and the cocoa and sugar confectionery industry.

The structure of the enterprises in the food industry is dominated by firms of micro and small scale: 75 per cent of all firms employ less than 10 people and 90 per cent of the companies have less than 50 employees and/or less than 10 million euro turnover.[1,2] Exceptions are the grain milling, sugar, beer and tobacco industries where medium to large companies represent more than 10 per cent of the total number of enterprises within their branch. Table 6.3 shows how the structure of the Serbian food industry compares with those in the EU-27 and with the other benchmark countries.

The average turnover per firm in the food and beverage industry differs strongly between Serbia, Bulgaria and Romania on the one hand and the EU-27 on the other hand: the average turnover per firm in EU27 is almost 3 times larger. The average size of the firm

[1] Most of the enterprises belonging to this category will be far from the threshold indicated, as the average turnover for these companies are no more than 250,000 euro.

[2] Enterprises can be classified according to their size, taking turnover, balance sheets and employment as key indicators. Definitions of small- and medium enterprises (SMEs) and large enterprises are similar in Serbia and the EU for the number of employees[2]: medium sized enterprises are with more than 50 but less than 250 employees (small <50; large has over 250 employees).

with more than 250 employees in the EU-27 is even 4 times larger than in the other countries (see Table 6.3).

Table 6.3. Number of enterprises and turnover of food and beverage industry in 2007.

	Total	Between 1 and 9 persons	Between 10 and 49 persons	Between 50 and 249 persons	250 or more persons
Number of firms					
EU-27	310,283	246,267	50,083	11,223	2,715
Bulgaria	5,300	3,501	1,328	417	54
Romania	9,993	6,896	2,382	579	136
Serbia	3,093	2,260	527	233	73
Turnover/firm in € million					
EU-27	3.01	0.25	2.74	22.32	178.40
Bulgaria	0.74	0.04	0.50	4.11	25.75
Romania	1.01	0.06	0.56	4.18	43.20
Serbia	1.65	0.13	1.15	5.87	39.02

Source: Eurostat SBS for EU-countries, SORS for Serbia.

Performance of the food industry

The utilization rate of the installed capacity of the Serbian food industry has been below 65 per cent in the period 2006 to 2009 (GoRS, 2011, Ch 20: 432). Industries at the higher end of the spectrum are the meat, sugar and dairy industry, while the processing of tea and coffee, beer and mineral waters and soft drinks are using the installed capacities 75-85 per cent in 2009. All other food subsector industries not mentioned have a low capacity utilization which points at low efficiency.

Statistics of the Serbian food industry are taken from GoRS, Ch. 20, that provides a selection of food manufactures at four digit NACE code. That selection counts almost 80 per cent of the total turnover of the food and drink sector, which has been around 5 billion euro in total in the years 2007 to 2009. Major subsectors in terms of turnover are the meat, dairy, grain milling and the beverages (beer, waters and soft drinks) industry (see Table A.1. in Annex for more detailed information).

Serbia's largest food branches are compared with the size and performance of those in the benchmark countries, Figure 6.4 shows that in the EU-27 meat, dairy, and bread and pastry industries have a higher share in the total food industry turnover than in Serbia, implying that the EU is more specialized in these branches than Serbia. On the other hand, Serbia's oil and fats, grain milling, cocoa and sugar confectionery, and beverage industries are relatively more important in Serbia than those in the benchmark countries. These figures on branches with a relative high share in the industry's total turnover points at their relatively strong position in their own country, yet do not indicate whether these branches are able to compete with those in the benchmark countries. Looking at the value added per industry (Figure 6.5), all Serbian selected industries score well compared to their competitors in the benchmark countries, generating similar percentages of value added as in other countries. However, as further detailed information on profitability, investments, product characteristics and other relevant indicators are lacking, this comparison of shares in turnover and value added is too small a basis for strong conclusions on the competitiveness of the Serbian food industry. Nevertheless, the typically small-scale structure of the industry, its overcapacity in nearly all branches and the generally too limited financial means to invest in modern technology (to improve both the efficiency of production and the quality of produce) reduces the future perspectives of the majority of food firms currently in business.

The sector performance indicators presented above also hide the fact that the industry is characterized by duality, with many (very) small and medium-sized companies and only a limited number of large-scale, modern firms. Studies focusing on individual supply chains (e.g. Van Berkum, 2009; Bozic *et al.*, 2010; Goss *et al.*, 2010; Gill *et al.*, 2010) point at small-scale and seasonal operation of plants, their low mechanization level, and the number of plants that do not comply with modern food safety and quality standards.[1] There are, though, also positive developments and modern and well equipped firms. In many cases foreign direct investment has acted as initiator of these developments, bringing in capital as well as technical and technological innovations. The food industry in Serbia has received significant foreign investment over the last decade. Largest amounts were linked to acquisitions and greenfield (new factories and/or stores) investments in the beverage industry and the food retail sector (Vladimirov, 2010), yet foreign investors also acquired (stakes in) companies, inter alia in the sugar, dairy and meat industries. The features of several of Serbia's major food industries are described in the next section, including an indication of their relative strengths and weaknesses.

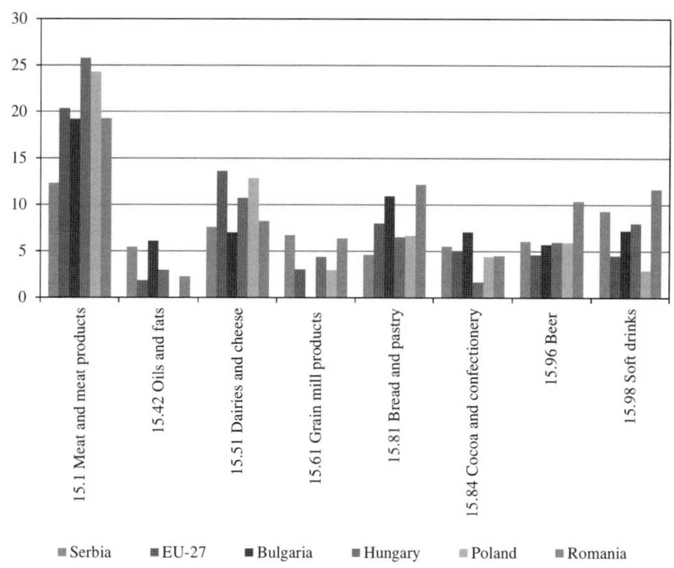

Figure 6.4. Share (per cent turnover) of food industries in total food and drinks industry in 2007 a).
Source: Eurostat SBS data; for Serbia: GoRS (2011).
a) Data for Hungary are from 2005.

[1] To illustrate the latter issue: to date only 6 slaughterhouses have an export certificate for the EU and 4 are allowed to export to CEFTA countries. Two dairy companies have a licence to export to the EU.

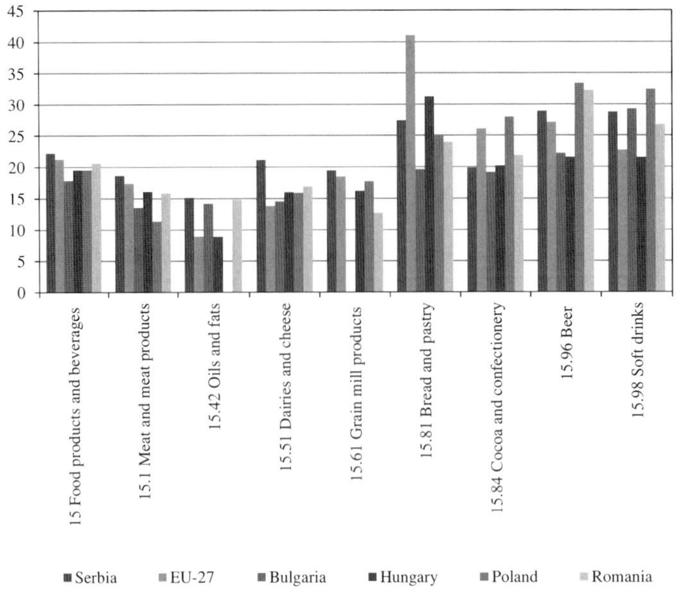

Figure 6.5. Value added as percentage of the turnover in 2007.
Source: Eurostat SBS data, Serbia: GoRS, 2011.
a) Data for Hungary are from 2005.

6.5 Characteristics of Serbia's major food industry branches

Milling and bakery industry

The large(r) factories in this sector date from the 1950s and have only partially restored facilities. Small(er) facilities mainly use old (previously used) equipment. Only about 20 per cent of the (mainly larger) companies invest in product innovation. Overall, the grain processing capacity is underused while about 80 per cent of utilized processing capacity is owned by 15 per cent of the (larger, more modern) companies (Bogdanov, 2010). In recent years multinational companies have shown an increasing interest in purchasing grain processing facilities. Foreign investors are expected to contribute to a further transformation of the sector, including further specialization and product differentiation, which will also lead to the closure of a large number of non-competitive companies. In the process of harmonization of legislation with EU standards, processing facilities may not be able to meet quality and food safety standards. Some of these facilities will be closed, and others will continue their operations by switching to the production of special flour and the finalization of semi-finished products.

Processing industries of industrial crops

The processing industry of vegetable oils and fats, sugar and oilseeds experienced a significant change and restructuring over the last decade. An estimated 90 per cent of facilities operating during the last five years has invested in new production lines, and about one-fifth of the enterprises has been completely modernized, using advanced technology (Bogdanov, 2010). Such facilities use approximately 80-100 per cent of their production capacity, yet some companies use only 20-40 per cent of their installed capacity. Almost all sugar beet, oilseeds and soybean processors apply food safety and quality standards (HACCP or ISO 22000), and more than half has a coordinated management system according to the ISO 9001 system. Opportunities for further growth of competitiveness may be in production of biofuel, and in strengthening cooperation with producers of raw materials to improve quality, enhance innovation and expanding assortment of production (especially rapeseed). Opportunities for foreign direct investment in this sector are relatively small, because the market and production is controlled by large multinational companies.

Sugar from sugar beet has become one of the most important export products of Serbia, primarily due to the preferential status of exports to the EU in 2000 with the introduction of Autonomous Trade Measures; the EU accounts for over 90 per cent of Serbia's sugar and sugar product exports (Bozic *et al.*, 2010) The harvested area increased from slightly over 40,000 ha in 2001 to almost 80,000 ha in 2007, yet has declined since then. Necessary inputs (fertilizers, seeds, etc.) have become much more expensive in recent years, affecting the farmers' profitability in sugar beet production. Three companies run the 8 sugar processing plants presently in operation, all of them of medium size. Fixed costs of sugar beet processing are high, reducing its competitiveness in the international market.

Over the last five years the confectionery industry has been upgraded; most companies use using appropriate, up-to-date technology. Capacity utilization rates vary across companies, with an estimated one third of the companies using up to 60 per cent of the installed capacity, but this is approximately the same percentage of those who use the full capacity. Improvement and innovation of the production program, including marketing in this sector, constantly takes place, given the strong competition within the CEFTA markets (Croatia, Macedonia).

Fruit and vegetables processing industry

This sector shows modern units next to traditional processing facilities using outdated technologies. Canned fruits and vegetables well as fruit juices are found in regional markets. Only the large(r) companies that are well integrated into the foreign market (buying raw materials locally and selling final products in foreign markets), as well as manufacturers with small factories that produce high-quality products, use their capacity up to 100 per cent. Opportunities for improvement are, among others, in the area of classification and standardization of products for fresh consumption for the foreign market and in the development of high-quality and/or organic products.

Meat sector

There is no available information about vertical integration in livestock sector, but it is estimated to cover about 15 per cent of the current market in Serbia. Many of the slaughterhouses have networks of dealers and agents travelling across the country purchasing livestock on an individual basis. Competition is great between these types of agents, who are often travelling well beyond their local market. Some of these dealers operate independently or on behalf of specific abattoirs. This system of organization of purchasing and transportation raises the question about fulfilling requirements of food chain information with some of these agents encountered, as well as with regard to animal welfare. The majority of animals are sold on a live weight basis. Price is negotiated between farmer and buyer, and may vary based on the end use of the animal. Generally per kilogramme price is applied, irrespective of the animal being sold for slaughter, or rearing or for breeding.

However, there are examples of contracted production, mainly between big producers and processors. The contracted production is increasing since the number of animals has become radically reduced. Contracts are usually initiated by slaughterhouses, which are faced with lack of raw material. A quality premium is defined by the percentage of meat in the carcass. Special bonuses are foreseen for the better bred animals in which the percentage of meat exceeds a certain limit, but other details are not defined in the contract agreements. Unfortunately in this system the farmer producing a quality animal is not rewarded for his efforts.

Poultry meat processing

As an important part of the Serbian meat sector, the poultry sector is characterized by little concentration, specialization, vertical integration and efficiency (Rodic *et al.*, 2011). Small-scale farms and backyard poultry production account for more than half of the poultry meat production and two third of egg production. Their produce is mainly for own consumption and sold on local markets for lower prices than commercial farms/firms, the latter facing tough competition on the domestic market. Rodic *et al.* (2011) assess the commercial (larger, specialized, vertically integrated) poultry companies as being uncompetitive at international markets due to high production costs per kg of meat (linked to low technical efficiency, such as feed conversion rates, low productivity at farm level, technical obsolescence of processing capacities and a rather limited range of final poultry products. The slaughtering industry is characterized by 50 per cent overcapacity. The authors also point at a serious lack of harmonized certification that limits the export to EU countries.

Dairy industry

ALC's assessment of the dairy industry points at the dominance of small scale plants, the rather traditional product assortment and the lack of an adequate laboratory network for milk testing, while the milk quality control system is not in full operation yet (ALC, 2010a). The majority of the companies would not be able to bear the costs of necessary investments to meet EU requirements on quality and hygiene. ALC expects a 40 per cent drop in the number of dairies within a 5 year period with a demise of most of the very small scale dairies. The EU's IPARD scheme will support the sector's modernisation and innovation investments, but this will go hand in hand with a significant structural change.

Fresh fruits and vegetables

The latter effect is expected in many more supply chains in the country, mainly because quality and food safety standards require investment which will not become profitable for many small-scale businesses. This probably also holds for the fresh fruit and vegetable (FFV) sector, which is contributing 30 per cent of total agri-food exports by value in 2009, and is widely perceived as one of the areas where Serbia could significantly increase its exports. Gorton *et al.* (2011) find that only a small number of export oriented producers in the Serbian FFV sector operate subject to extensive private regulation while the majority of

FFV farmers supplying the local market occupy regulatory voids, immune to both public and private control. At present the majority of FFV producers in Serbia serve the domestic market with little pressure to adopt private standards. However, the authors point at a situation in which the rapid growth of supermarket chains in the country make private standards more salient, and in which, if small-scale FFV producers were unable to upgrade, a significant shake out of the supply base would be likely. Those remaining in farming but excluded from supplying supermarkets face being condemned, in the long-run, to supplying only increasingly low value added and declining wholesale markets.

6.6 Primary agriculture

The most recent data on the Serbian farm structure is from 2002 (see also chapter 4). According to that data the agricultural structure in Serbia is made of many small scale farms and only a few large holdings: farms above 20 ha represent less than 1 per cent of the holdings and use 7 per cent of the agricultural area. However, the 2002 census data of the Serbian statistical office refers to private family farms only, without registering agricultural enterprises and cooperatives. This old and incomplete data inhibits a comparison in farm structures between Serbia and the benchmark countries, although generally the structural characteristics pictured by the 2002 census have not changed much over time; the dominancy of small farms in Serbia is still a reality in both crop and livestock production. Popovic (2009), among others, shows that over 50 per cent of the dairy cows are on almost 82 per cent of the farms with 1 to 3 cows. The largest 92 farms (0.14 per cent of all dairy farms) with over 50 cows possess only 4.2 per cent of all dairy cows. Rodić *et al.* (2011) indicates that small-scale and backyard poultry production accounts for over 50 per cent of the broiler production and two thirds of the egg production.

6.7 Suppliers in the upstream industries

The Serbian agricultural sector is basically a low-input production system, using relatively few agrochemicals and compound feed for increasing yields. The reason is the relatively high prices of the inputs for the majority of the (small-scale) farmers, which have limited financial resources and cannot afford to buy these inputs. Consequently Serbia has very low processing capacity for agro-chemicals (fertilizers, pesticides). Major companies active in this area are traders, importing and exporting the ingredients for fertilizers, while a few firms produce and distribute fertilizers and/or pesticides domestically. Seeds for planting are largely generated from previous crops, which results in a rather limited market for commercially certified seeds for maize, wheat and/or vegetables (USDA/FAS, 2011). Major players in seed production are two semi-state owned institutes controlling over 60 per cent of the maize seed market in Serbia, with some foreign suppliers active in this and the vegetable seed markets.

As livestock farming accounts for one third in the agricultural output, animal feed is an important input to this sector. Small scale feed mills characterize the Serbian animal feed industry: 86 per cent of the total mills produce altogether 20,000 tonnes per year, whereas the total production in 2004 was about 500,000 tonnes (Stankovic, 2009). The main raw material for processed animal feed is maize, accounting for more than 50 per cent of the total volume, plus soybean meal that is imported. Stanokovic analysed 9 feed mills, representing 20 per cent of the total Serbian feed mill capacity and 16 per cent of the total employment in this sector. The capacity utilization, though, is very low across the board, as most of the enterprises in the sample for this study have a capacity utilization (far) below 40 per cent.

6.8 Relationships between farmers and markets

Linkages between farmers, traders and processors can be either through the spot market or through contracts. Contract farming has not been widespread in most agricultural subsectors in Serbia. This is due to the small-scale features of both the primary agriculture and a large part of the processing industry. For fresh products, such as fruit and vegetables, the large group of small-scale producers prefer the farmers' markets as outlet for their products. ALC (2010b) points at the tendency that farmers in these subsectors tend to prefer the speculative nature of the spot markets, and even when they have a contract with processor, farmers tend to sell to the first buyers approaching them with cash during harvest time. Processors, then, face difficulties to secure collection of adequate volumes of quality products to feed their plants. In order to have consistent supply of raw material, owners of the main processing units are investing in land to be able to directly produce the raw material they need.

On the other hand vertical coordination through contract farming occurs in some sectors when agricultural products have to go through a processing stage. Examples are sugar beet growing, which is fully under contract with the processing factories, and industrial crops (sunflower, soybean, tobacco). In the livestock production, contracts for delivery are largely set only between the larger producers and processors (ALC, 2010a). Vertical integration is only occurring to some extent in the poultry and pig meat production (Gill *et al.*, 2010), where a number of large meat processing plants have moved upstream in the supply chain and maintain their own pig and cattle feeding lots. They do so to ensure continuity of supply of often specific types of animals for specific processing lines within their meat factories. A number of larger units are owned in this manner.

A large part of the primary agricultural sector is excluded from the commercially oriented agri-food supply chain, because due to its small scale of production it cannot provide the industry with uniform quality of raw material in sufficient volumes and efficient ways. Horizontal integration -- farmers cooperating in producers' organizations organizing the marketing of their products -- could be an instrument to link the small-scale farmer to processors, but to date there has not been much interest in establishing producer groups (e.g. Popovic, 2009).

With the emergence of the modern retail chains, suppliers of fresh and processed food products are more and more subject to their requirements. Shops and supermarkets obtain their products through contracts with processors, in which the terms of delivery such as price, volumes, quality and timely delivery are specified. In this context, scale matters: the larger the business, the greater the negotiating power.

In summary, farmers and processors trade mainly through spot markets. Yet, the processors increasingly face the terms and requirements set by the modern retail chains, that use contracts as a major governance structure. This development may require a further integration of the primary sector into the agri-food supply chain in due time, in order for it to remain the main source for the food processing and retail sectors in Serbia.

6.9 Conclusions

The structural features of the agri-food supply chain in Serbia point to a dual structure in every part of the chain. A large number of small to very small farms and firms dominate both the primary farm sector as well as the upstream and downstream industries. Next to these there are only a few medium to large scale operators. At the primary level, farm structures are highly fragmented, with a large part of the farm holdings being (semi-) subsistence. The structure of the food processing industry varies across branches. Nevertheless, only a few companies account for a major share of turnover. Market power, however, does not seem to exist, although hard evidence to verify this is not available.

Home consumption and direct sales of food (the informal economy) is highly significant in Serbia. This implies that a large part of the agricultural sector is not integrated in the agri-food supply chain, either by contracting or by (any other means of) sustainable commercial relationships. This is the large group of (very) small-scale farmers. Next to that, there is a large number of small processing units. Their position is critical as they have to comply with the requirements of an increasingly demanding retail chain. These requirements are driven by the coinciding forces of increasing consumer awareness and public policies aiming at establishing the legal framework for food safety and quality standards. Even the larger enterprises have difficulties in adapting to the European requirements, given the limited number of licenses for export to the EU within the meat and dairy sectors. Foreign investment in the retail sector as well as in the food processing industry may contribute to enhancing competitiveness, bringing in capital as well as technical and technological innovations.

The food chain in Serbia is mainly governed by free market principles with arm's length price formation. This holds especially for the relationship between the primary and the processing levels of the supply chain, although several branches of the processing industry work with contractual agreements for the supply of raw materials. The modern retail chain sector in Serbia demands ever more stringent guarantees on quality and food safety at competitive prices. These guarantees are enforced by contracts. A major challenge for the farming and processing sector is, then, to meet the growing number of requirements and standards all along the food supply chain. Failure to meet such requirements may result in increasing import penetration.

Literature and websites

Agri-Livestock Consultants (ALC), *Dairy sector study for the IPARD programme, Republic of Serbia.* Final Report, September. Framework contract EuropeAid/127054/SV/Multi., 2010a.

Agri-Livestock Consultants (ALC), *Fruit and Vegetable sector study for the IPARD programme, Republic of Serbia.* Final Report, June. Framework contract EuropeAid/127054/SV/Multi., 2010b.

Beneke, J. 'Consumer perceptions of private label brands within the retail grocery sector of South Africa.' In: *African Journal of Business Management* 4 (2010), pp. 203-220.

Bogdanov, N., *Small rural households in Serbia and rural non-farm economy.* Belgrade, UNDP, 2007.

Bogdanov, N., *Effects of Serbian Integration in the EU – Agriculture.* FEFA, Belgrade, 2010.

Bogdanov, N. and M. Petrovic, *Financial poverty and material deprivation in rural Serbia. Poverty and Social Inclusion in the Western Balkans.* Brussels, 2010.

Bord Bia, *Serbia Market Overview*. Frankfurt, 2008. Available at: www.bordbia.ie/eventsnews/ConferencePresentations/FoodDrinksIndustryDayCountryOverviews/SerbiaMarket20Overview.pdf. Accessed 24 April 2011.

Bozic, D., P. Muncan and N. Bogdanov, 'Economic characteristics of sugar sector in Serbia.' In: *Research Journal of Agricultural Science* 42 (2010) 3.

Eurostat, *Household Budget Survey*, Eurostat, Luxemburg.

Eurostat, *Structural Business Statistics*, Eurostat, Luxemburg

FAO, *FAOStat*. FAO, Rome. www.fao.org/faostat

Gill, S., V. Bogdamovic, C. Longo and J. Whelan, *Meat sector study for the IPARD programme*. Hemel Hempstead, HTSPE LTD (UK), 2010.

GoRS, 'Answers to the European Commission's questionnaire.' In: *Chapter 20: Enterprise and industrial policy*. Belgrade, 2011.

Gorton, M., V. Zaric, P. Lowe and S. Quarrie, 'Public and private agri-environmental regulation in post-socialist economies: Evidence from the Serbian fresh fruit and vegetable sector.' In: *Journal of Rural Studies* 27 (2011), pp. 144-152.

Goss, S., J. Howells, M. Luijks, S. Hopic, R. Popovic and Z. Janovic, *Dairy sector study for the IPARD programme Republic of Serbia*. Agri-Livestock Consultants Ltd, Chester, UK, 2010.

Interbrand, *Private brands. A global guide to the rise of private label brands*. Available at: www.interbrand.com. Accessed 18 April, 2011.

Koncar, J., G. Vukmirovic and Z.P. Katai, 'Private label development in the Republic of Serbia.' In: *Business and Economic Horizons* 3 (2010), pp. 105-110.

LSMS, Living Standards Measurement Survey 2007. Available at www.worldbank.sr

Popovic, R., 'Effects of market structure changes on dairy supply chain in Serbia.' In: *IFAMA Forum and Symposium Proceedings*, Budapest, June 20-23, 2009.

RBR, Delhaize to buy Serbia's Delta Maxi for $1.29bn [Online]. Available at: http://hypermarketsandsupermarkets.retail-business-review.com/news/belgian-supermarket-group-delhaize-to-buy-serbias-delta-maxi-for-129bn-040311. Accessed 18 April, 2011.

Rodic, V., L. Peric, Z. Pavloski and N. Milosevic, 'Improving the poultry sector in Serbia: major economic constraints and opportunities'. In: *World's Poultry Science Journal* 66 (2011), pp. 241-250.

Stankovic, V., 'Influence of organizational and economic factors on business results of feed mills in Serbia.' In: *Journal of Agricultural Sciences* 54 (2009), pp. 166-178.

Statistical Office of the Republic of Serbia, *Statistical Yearbook*, various years. SORS, Belgrade.

USDA/FAS, *Serbia. Grain and feed annual report*, GAIN report number; RB1107. 2011.

Van Berkum, S., 'Dairy chain competitiveness in EU's new member states, candidate and potential candidate countries.' In: *Agrarwirtschaft* 58 (2009), pp. 314-323.

Vladimirov, S., *Foreign direct investment in Serbia, with special attention to Austrian FDI in Serbia*. Master thesis. Universität Wien, 2010.

World Bank, *World Development Indicators*, World Bank. Washington D.C 2011.

Annex Table 1 Turnover (in €1,000), number of enterprizes, employment and capacity utilization (%) of the Serbian food and drink industry in 2009

NACE Rev. 1	Turnover	Number of enterprises	Employment	Capacity utilization
Food and drinks sector				
15 Manufacture of food products and beverages	4,925,748	3337	73,734	59.1
15.1 Production, processing and preservation of meat and meat products	590,545	409	11,473	62.1
15.42 Manufacture of refined oils and fats	217,369	12	1,785	30.0
15.43 Manufacture of margarine and similar edible fats	1,249	1	5	41.6
15.51 Operation of dairies and cheese making	418,538	179	4,905	72.1
15.52 Manufacture of ice cream	87,281	23	1,418	78.2
15.61 Manufacture of grain mill products	338,194	232	6,165	51.5
15.62 Manufacture of starches and starch products	11,629	5	443	44.9
15.81 Man. bread; manufacture of fresh pastry goods and cakes	223,583	598	9,168	59.7
15.82 Man. of rusks and biscuits or preserved pastry goods and cakes	205,718	57	3,969	57.3
15.83 Manufacture of sugar	175,573	15	1,265	97.7
15.84 Manufacture of cocoa; chocolate and sugar confectionery	226,904	83	4,209	54.0
15.85 Manufacture of macaroni, noodles, couscous and farinaceous products	11,119	48	394	49.6
15.86 Processing of tea and coffee	255,203	144	1,810	80.1
15.87 Manufacture of condiments and seasonings	26,825	77	484	32.1
15.88 Man. of homogenized food preparations and dietetic food	14,808	68	374	10.4
15.89 Manufacture of other food products n.e.c.	186,265	136	3,228	63.1
15.91 Manufacture of distilled potable alcoholic beverages	63,382	81	1,216	9.7
15.93 Manufacture of wines	82,101	73	1,756	25.0
15.94 Manufacture of cider and other fruit wines	4	2	1	59.3
15.95 Manufacture of other non-distilled fermented beverages		1	-	
15.96 Manufacture of beer	290,718	28	3,140	63.9
15.97 Manufacture of malt	21,032	1	53	84.2
15.98 Production of mineral waters and soft drinks	386,521	163	5,134	73.5
16 Manufacture of tobacco products	287,448	11	1,681	53.7

Source: GoRS, chapter 20 (2011).

Chapter 7

7 Overview of agricultural, rural and structural policies

7.1 Introduction

This chapter explains the principles and mechanisms of the Serbian agricultural market and rural development policies, and discusses recent reforms. By doing so this chapter provides insights into the consistency of the Serbian with the EU agricultural, rural and structural policies.

Serbia (i.e. former Yugoslavia) has a long history of agricultural policy research carried out by academics and research institutes. This research was mainly descriptive, focused on social and political aspects of the sector, with a strong emphasis on political promotion and glorification of the agricultural model based on big socialistic vertically integrated companies. Objective quantitative analysis of policy instruments did not exist until the end of the 1970s. During the 1980s, under the influence of co-operation with OECD and other international institutions, the federal administration has started to make analyses of budget allocations in accordance with the AMS (Aggregate Measurement of Support) system. Because of incomplete data and the difficulty to separate the parts of budgets dedicated to each republic, these sources are not used for this study. The general concept of agricultural policy in the period up to the beginning of 2000 has been presented in Chapter 3. The basic instruments of agrarian policy were high border protection, price support (guaranteed prices for about 20 strategic products, premiums and input subsidies) and subsidized loans for purchase of machinery and for investment in livestock production.

In this chapter we analyse the main elements of agrarian policy reforms Serbia applied since 2000, including strategic objectives, the amount of budget allocation and the implemented mechanisms and main measures. The quantitative analysis uses figures for the period 2004-2009. During the period 2000-2004 support to agriculture was financed from both the republican and federal budgets (FRY), and because of this it is impossible to derive reliable and comparable data. The support measures are classified based on the OECD systematization. The amount of support shown in the tables refers only to the national agricultural budget. Municipalities, donors and other sources are not taken into account.

7.2 Main elements of recent reforms

Systemic and structural reform of the agrarian sector of the Republic of Serbia began after the political changes in 2000. Since then, agricultural and rural policy in Serbia has gone through several phases.

Many years of overlapping of competencies in the field of agriculture between the Serbian and federal institutions (FRY -- Federal Republic of Yugoslavia) held back the transformation of the agricultural policy framework and the establishment of institutions necessary for policy implementation. Namely, the process of transferring powers in the field of agriculture from the federal to the republic level was not completed before 2004. Before the State Union of Serbia and Montenegro was formed in 2003, the ministries of agriculture on both the federal and the republican levels operated parallel. With the establishment of the State Union of Serbia and Montenegro, the Ministry of Agriculture at the federal level was abolished, and part of its responsibilities (sanitary controls, export incentives, etc.) was distributed between the Federal Ministry of Economy and Internal Trade (which continued to function) and the Ministry of Agriculture of the Republic of Serbia. After the separation of Serbia and Montenegro, all policy responsibilities for the sector were taken over by the MAFWM. Another reason for slow institutional transformation may be attributed to frequent changes of the administration in MAFWM. The lack of continuous and consistent policy, i.e. clearly-defined priorities and the measures and mechanisms of policy implementation, led to low efficiency and to the overall results of the sector far below expected levels.

The agrarian policy of Serbia in the early 2000s was aimed at solving the urgent problems of agriculture and rehabilitation of the sector. Returning to the EU market after decades of isolation and asymmetric trade preferences had a crucial impact and resulted in increasing exports, especially of sugar and vegetable oil. Therefore, the production of industrial plants was encouraged by direct payments per hectare. During that period, specific land policy was introduced which included incentives for farmland renting. In the period 2000-2003 the supervision and operation of inspection services in those parts of the sector that were most affected by the grey economy was significantly improved (Bogdanov and Bozic, 2010).

From 2004-2007, a strong turn was made in the strategic determination and implementation mechanisms in the agricultural policy of Serbia compared to the previous period. The reformed agricultural policy was intended to increase the competitiveness of commercial family farms. In terms of implementation mechanisms, the agricultural policy focused on encouraging investments on farms. The system of direct payments was established, the support system for investments intended for rural development provided credit with more favourable conditions than market loans, renting of agricultural land was encouraged and the registration of agricultural holdings was initiated.

From 2007, the implementation of agricultural policy has been constantly changing. Programmes and regulations were changed and/or abolished several times during the year, and payments to the users were delayed, which contributed to the creation of an unstable and unfavourable economic environment. Additionally, there was no defined model for monitoring and reporting agricultural policy implementation and related budgetary expenditures, thereby making it difficult to assess the situation. Since 2007, direct payments per hectare of sown area/head of livestock are an important element of the budgetary support scheme, and accompanied by support to investments intended for rural development and modernization of agriculture through granting credits under subsidized conditions. The year 2010 is dominated by the model of direct payments, and in comparison to the previous period the scope of incentives intended for the implementation of the rural development policy has been increased.

Legislative work in the field of agriculture and rural development was intensified in the period of 2008-2009. The required basis for faster progress in aligning with the requirements of thc *acquis communautaire* was provided in areas of key importance for the agricultural sector of the Republic of Serbia, which will contribute to increasing its competitiveness in the international market.[1] The Law on Agriculture and Rural Development, as the framework law of the agrarian sector and rural development, sets the aims of agrarian policy and defines the types of incentives, registry of farm holdings, types of holdings and a range of other solutions pertaining to rendering operational the policy in the field of agriculture and rural development. The structure of the Department for Rural Development was aligned with the requirements for the realization of IPA component V - IPARD (Instrument for Pre-accession Assistance - Rural Development). The Directorate for Agrarian Payments, responsible for payment of agricultural subsidies, was established and the reforms of organization and scope of activities of agricultural extension services was completed. With a view to improving rural development policy coordination, an inter-ministerial body, the National Council for Rural Development, was established towards the end of 2010; it has the aim of coordinating rural development measures and programmes with other ministries and strengthening regional ties, liaising with local communities, ensuring two-way information flow and building partnerships at the local level (GoRS, 2011b).

7.3 General agricultural policy framework

'The FRY long-term agrarian development policy' was prepared in 1995, and by 2005 it was the only officially adopted document about agricultural policy. This document consisted of three parts: 1) strategic goals of agrarian development, 2) development courses and structural changes and 3) long-term macroeconomic instruments for agrarian policy. Six strategic goals of the agrarian development have been established, based on the socio-economic structure and assessment of the agricultural resources. The starting point is the significance of the agricultural dynamic development for the country's economic, social, political and security stability, taking into consideration economic globalization and liberalization processes worldwide. This strategic document emphasized the role of agriculture as an integration factor in strengthening regional cooperation (South East European Stability Pact), as well as for reinforcing the newly established relation between the federation and the republic relative to their respective jurisdictions (Milanovic, 2006).

In 2005, the government adopted the Agriculture Development Strategy. The reformed agricultural policy was intended to increase the competitiveness of commercial family farms. In terms of implementation mechanisms, agricultural policy focused on encouraging investments. The Strategy includes the following objectives: 1) build a sustainable and efficient agricultural sector that can compete in the world market, and thereby contribute to GDP growth; 2) provide food products which satisfy consumers' needs in terms of quality and safety; 3) ensure the living standard for people who are dependent on the agricultural industry and are incapable of keeping up with economic reforms; 4) ensure support for sustainable rural development; 5) protect the environment from the negative effects of

[1] The key laws passed include: the Law on Agriculture and Rural Development, the Law on Advisory and Extension Services in Agriculture, the Law on Wine, the Food Safety Law, the Law on Animal Welfare, the Law on Plant Health, the Law on Livestock, the Law on Agricultural Land, the Law on Organic Production and the Law on Public Warehouses for Agricultural Products.

agricultural production; 6) prepare the Serbian agricultural industry for EU integration; and 7) adapt domestic support and trade policy to WTO standards.

Meanwhile, the financial *crisis* and economic downturn in 2008 and 2009 required the redefinition of agricultural policy measures, taking into account the state budget deficit and the importance of agriculture to the macroeconomic indicators of the country. Two documents were adopted: National Agricultural Programme of the Republic of Serbia for 2010-2013 and National Rural Development Programme of the Republic of Serbia for 2011-2013. Both programmes define long-term paths of development of agriculture and rural areas, and show the government's orientation towards developing the sustainable economy concept promoting environmental protection and sustainable management of natural resources.

Besides the two Programmes, a number of other strategic documents were adopted in the period 2001-10 relating (directly or indirectly) to certain aspects of rural development. It is indicative that only the recent strategic documents explicitly use the term rural development and devote adequate interest to rural areas within the issues they consider. Strategies adopted at the beginning of the transitional period did not explicitly point out rural areas, their potential and limitations or structural problems, but they were considered within topics such as: agriculture; underdeveloped regions/areas, population aging and migration, environmental protection, development of SMEs; rural/farm tourism, etc. (Bogdanov, 2008).

7.4 Budgetary outlays to agriculture

The allocation and use of budgetary funds for agriculture and rural development is carried out according to a special act passed by the government, on the proposal by MAFWM, in line with the Budget Law of the Republic of Serbia. Based on the approved funds, MAFWM adopts the programme of allocation and use of the subsidy system in the fields of agriculture, forestry and water management for the current year. The funds and activities stipulated by the programme are made operational through:

- Regulations adopted by the Government based on MAFWM proposals.
- The open call for tenders meant for programme activities which are eligible for co-financing by the users of the funds.
- Programmes designed to finance actions of general interest for which MAFWM provides the scope and programme of activities in the current year (extension service, etc.).
- Direct funding, based on the beneficiary's request. This is a method of funding a wide range of initiatives related to improving the overall capabilities of the ministry and other institutions (publications, promotional materials, conferences, commissions, etc.).

During the period 2001-2010 large fluctuations in level and structure of support occurred (Bogdanov *et al.*, 2010). Quantitative analysis of the agricultural budget clearly indicates a significant reduction in 2005 compared to 2004 (Table 7.1). After 2005, budgetary support again increased, and peaked in 2008. Of the total funds spent in 2008, 24.6 million euro were obligations from 2007. Bearing this in mind, it may be concluded that budget growth was more modest. However, a large delay in the distribution of about 12 per cent of funds for incentives is not negligible from the standpoint of the users of these resources.

Table 7.1. Budgetary support to agriculture (in million euro) [1]

	2004	2005	2006	2007	2008	2009
1.Market and direct producer support measures	190.794	102.593	99.051	159.008	222.671	112.253
1a. Market support measures	60.847	10.223	22.384	24.703	12.208	8.585
- Export subsidies	15.522	10.223	10.723	12.217	12.208	8.585
- Market intervention	45.325		0.734	12.486		
- Other			10.927			
1b.Direct payments and variable input subsidies	129.947	92.370	76.667	134.305	210.463	103.668
1c. Direct payments to producers	85.419	62.030	32.772	28.307	21.358	10.676
1d. Direct payments based on output (price aids)	85.419	53.180	32.772	27.058	14.883	5.809
1e. Direct payments based on current area/animal heads		8.850		1.248	6.475	4.867
1f. Variable input subsidies	44.528	30.340	43.894	105.998	189.105	92.992
2. Structural and rural development measures	42.520	40.405	81.176	32.447	26.212	15.631
2a. Improving the competitiveness of the agricultural sector	42.472	37.142	79.569	20.215	21.357	12.336
2b. Improving the environment and the countryside	0.049	2.069	0.389	0.581	0.377	
2c. Supporting rural economy and population		1.193	1.218	8.757	1.704	0.500
2d. Miscellaneous rural development measures				2.893	2.774	2.795
3. General measures related to agriculture	5.010	5.881	3.744	4.248	4.494	5.717
4. Miscellaneous agricultural policy measures	10.590	1.512		5.397	24.647	30.022
5. Budgetary expenditure for agri-food sector and rural areas	248.915	150.390	183.970	201.099	278.024	163.623

Source: own calculation based on internal MAFWM data.

Between 2004 and 2006 the budgetary support for market and direct producer support measures rapidly decreased, and they increased again after 2007. The reason for the reduction in 2005 may be found in the reduction of the total budget (there are no important

[1] The analysis used the records of funds spent to support agriculture and rural development by individual budget lines, provided by MAFWM. Source of data for the period 2004-2006 was the accounting centre of the Ministry. For the year 2006 the data from the MAFWM financial service have also been provided, and for 2007-2009 only the data from the MAFWM financial service.

differences in the structure of the budget), while in 2006 it was a consequence of the large increase in the execution of second pillar measures (introduction of support to non-commercial farms). Aside from 2006, the share of the market-price policy measures in the total budget of all three years was quite similar (from 73 to 80 per cent). On the other hand, looking at the composition of those support measures, radical changes are noticeable in their amounts.

The main problem of such frequent and radical changes in types and forms of direct support to producers is that these changes are generally connected with a re-distribution of funds among the farmers and did not create a stable economic environment for producers.

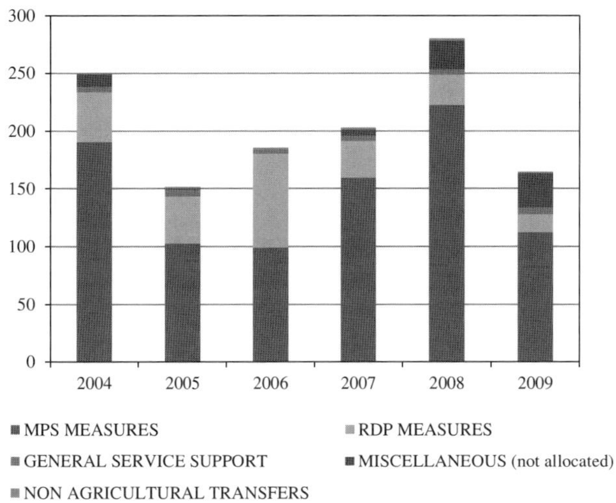

Figure 7.1. Break down of total budgetary sources (million euro).
Source: Own calculation based on data of MAFWM.

Within the structural and rural development policies, support to non-commercial farms was implemented in 2006 (when the volume of funds for this purpose was quite significant) and in 2008. Excluding the funds intended for this measure, it can be seen that the majority of funds was spent in the form of support to farm investments. The amount of these funds was greatly reduced in 2007-2008. Actually, with the change of administration in MAFWM in the middle of 2007, support to rural development was reduced to a minimum. Most of the funds were granted through investment support or input subsidies (in order to alleviate the effects of drought) (see also Bogdanov *et al.*, 2008).

Funds for supporting various services in the observed period were relatively small. Such a small amount of funds to support the development of service providers may be partly a result of the fact that they were financed from donor projects and other funds.

7.4.1 Market and direct producer support measures

The aim of market policy is to guide agricultural production and stabilize markets. The EU market of the most important agricultural products is regulated by a Common Market Organization (CMO) system which encompasses a wide range of possible interventions. Even though the CMOs for all 21 products have similar composition, there are differences in organizational detail. They provide guarantees which vary according to the particular economic and agricultural characteristics of the products and are grouped in two main headings (Massot Martí, 2010):

- The measures that regulate the market within the EU borders, the 'internal market' heading which includes common price systems, measures to control supply and stocks, measures to support production and measures for the organization or regulation of the market;
- an 'external market' heading (related to both imports and exports), with a system of protection against third countries, rules for the administration of tariff quotas, safeguard clauses and export subsidies.

Changes in the aims and means of organizing the markets since 1992 have changed the design of the CMOs, which now could be classified in four categories according to the price mechanisms of the single CMO used, on the one hand, and the direct support regimes established for the same products by Regulation (EC) No 73/2009, on the other hand (Table 7.2). Direct support is defined as recurring non-market transfers to farmers whether they are production related or not. Stimulated by the switch away from price support in the early 1990s (the MacSharry reform), direct income payments have now become the central policy tool in EU agriculture. More than two thirds of the CAP budget is now accounted for by these 'single' payments, which are largely decoupled from production. To receive these payments, farmers are required to comply with minimal 'cross-compliance' standards related to food safety, hygiene, the environment, animal welfare and land management (Silvis and Lapperre, 2011; Jongeneel and Brand, 2011).

Table 7.2. Products listed by type of intervention price (Regulation (EC) No 1234/2007 - single CMO) and direct support system (Regulation (EC) No 73/2009) following the Health Check.

Single CMO / Direct support	Products with CMOs at common prices with automatic intervention	Products with CMOs at common prices and conditional intervention	Products without regulated prices and direct aid only	Products with customs CMOs or without CMOs
Products with decoupled aid (SFP)	Cereals (except rice and rye)	Beef and veal, exceptional measures to support markets in cases of animal diseases for all livestock production (Article 44 single CMO)	Rye, sheep meat and goat meat, oilseeds, processed fruit and vegetables, hops, flax and hemp, dried fodder, olive groves, legumes, tobacco (with restructuring measures).	
Products with decoupled aid (SFP) and/or aid coupled to production (usually		Wine products (with distillation), fresh fruit and vegetables (with withdrawals)	Suckler cow premium, ewe and goat premium, starch potatoes (2012), protein crops	

Table 7.2. Products listed by type of intervention price (Regulation (EC) No 1234/2007 - single CMO) and direct support system (Regulation (EC) No 73/2009) following the Health Check.

Single CMO / Direct support	Products with CMOs at common prices with automatic intervention	Products with CMOs at common prices and conditional intervention	Products without regulated prices and direct aid only	Products with customs CMOs or without CMOs
optional)			(2011), nuts (2011), seeds (2011)	
Products with direct aid coupled to production	Sugar (reference price and national production quotas), rice (until 2011), dairy products (with national production quotas)		POSEI, bananas, cotton (with restructuring measures), sugar beet (2014), nuts (national aid), fruit and vegetables, including red fruits (in EU-12)	Poultry and eggs, pig meat (health crises - Article 44 single CMO)
Products without direct support				Processed agricultural products (PAPs), live plants and cut flowers, some fresh fruit and vegetables, potatoes, ethyl alcohol, silkworms

Source: Massot Martí (2010).

7.4.2 Market support measures

Among market support measures, import protection and export subsidies have been the most widely used mechanisms for supporting agriculture in Serbia. Market intervention purchases have only been implemented occasionally; there is no Agency for market interventions in Serbia in the way the system works in the EU (see also section 3.7).

Reforming Serbian foreign trade regulations started in 2000 with Amendments to the Foreign Trade Act and Foreign Currency Act. In 2001, the Serbian Government, in cooperation with the International Monetary Fund and the World Bank, made further steps towards liberalizing foreign trade; quantitative restrictions for a majority of products were abolished, a number of products were exempted from the export quota regime and the Customs Tariff Act was adopted. Nevertheless, the share of agricultural products in the group of products with maximum customs protection remained high, with maximum tariff rates being introduced for strategic agricultural products. Generally, in determining the customs charges for agricultural and food products, the principle was followed that the products which were not produced in the country or not produced in sufficient quantities (breeding animals, additives, cocoa, citrus fruits, citrus fruit concentrates), were charged lower rates (1 per cent and 5 per cent). Higher rates (10 per cent and 20 per cent) were anticipated for products that may also be used as intermediates and are not produced in sufficient quantities in the country, as well as for the commodities for which import was not quantitatively restricted before (see also chapter 8).

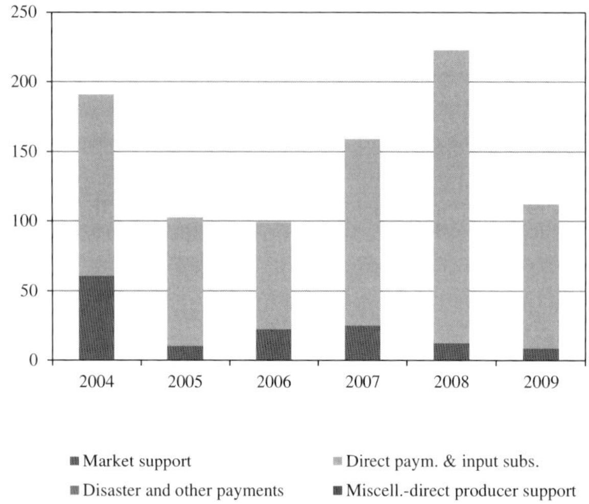

Figure 7.2. Break down of market and direct producers support measures (million euro).
Source: Own calculation based on internal data of MAFWM.

Export subsidies were applied after 2000 for a wide range of agri-food products (meat and meat preparations, dairy products, fruits and preparations of fruits and vegetables, beverages). Export subsidies absorb most of the market support funds (approximately 12 million euro from 2004-2009, on average), and were practically the only market support measure implemented in 2008 and 2009.

The objective of the measure is improvement and promotion of domestic production, relief from local surpluses and finding new markets for products that are not competitive on foreign markets. Those who are entitled to these funds are legal entities and entrepreneurs operating as exporters; and since 2010 natural persons, such as holders of commercial family farms, are also entitled to these funds for export of their own agricultural and food products. Since 2006, the right to use the sources according to this regulation does not apply to exporters who export commodities of domestic origin on the territory of the countries signatories of CEFTA 2006. The basis for calculation of funds for subsidies is the dinar equivalent to foreign exchange collected for export of goods after deduction of foreign costs (bank costs, fees, transportation, insurance, etc.) (GoRS, 2011a).

Direct producer support measures

Producer (production) support is the dominant form of budgetary support to agriculture in Serbia. In recent years this support varied significantly in terms of content, value and the method of implementation. Support programs were often changed, even within one production year, which created an uncertain business environment for producers. Since 2000, direct support schemes were implemented for the production of wheat, industrial crops and milk, and since 2007 also for young bulls and dairy cows. As of 2007, the crop premiums were abolished (except the tobacco premium) and replaced by input subsidies (see Table 7.3 below).

Subsidizing of the production and storage of wheat

The premium for wheat production was introduced in 2004 and the support scheme changed several times during that year. The original solution anticipated subsidizing the wheat producers for the quantities stored at registered warehouses. The anticipated amount of premium was 13.78 euro/t, with a supplementary payment of 6.89 euro/t if the stored wheat was immediately sold at a price of 89.57 euro/t, or at 96.46 euro/t paid in three instalments. An even more controversial solution was implemented in 2005 when the market price of wheat at harvest time was extremely low; because of this, MAFWM adopted the regulation of producer subsidies amounting to 60.31 euro/ha. However, this support was paid to producers who had signed a form stating that they intended to produce wheat in 2005, and not to those who had produced wheat that year and had really been hit by low prices.

In 2010 a new measure has been introduced with the aim to stimulate farmers to store their products in the public warehouses. This measure ensures that the goods are not sold immediately after harvest when prices are generally lower, but are sold when their price in the market is the most favourable. Incentives are implemented for wheat, durum wheat and maize for the area up to 100 ha per farm.

Industrial crops

Premiums for the production of soybean, sunflower and sugar beet were introduced in 2001. The aim of these measures was to stimulate production of industrial crops in order to help the food industry to recover and to fill the sugar quota approved by the EU. Premiums for the production of sugar beet have been paid according to the sown area contracted with sugar factories. In 2004, support for sugar beet was 41.34 euro/ha but in 2005 it fell again to 12.06 euro/ha. Premiums for soybean and sunflower production were granted for the quantities delivered to processors within contracts. Besides mercantile soybean, the MAFWM 2004 Program also included the production of certified soybean seed (with the same premium amount) in order to improve quality and combat the illegal imports of GMO seeds. In 2006, after this production had been increased and stabilized and the privatization of the food industry ended, direct payments for industrial crops were abolished.

A premium for oil pumpkins was first introduced in 2004 (110.24 euro/t). In 2005, the premium was reduced (72.37 euro/t) and then in 2006 changed to an area payment (83.15 euro/ha) granted to registered farms that declared themselves as commercial producers. A premium for rapeseed was granted only in 2006 in order to meet the needs of the first biodiesel production plant that was opened in Serbia. This premium was also granted in the form of area payments (83.15 euro/ha) to registered farms that declared themselves as commercial producers.

The objective of the premium for tobacco was to maintain or increase the production areas, improve the quality of tobacco and increase the competitiveness of producers. Until 2005, the requests were submitted to the Ministry of Agriculture, Forestry and Water Management, through the organizers of production, and since 2006, to the Ministry of Finance (Treasury Administration). Up to 2005, premiums for tobacco were granted for the contracted and delivered product, depending on variety and class. The highest premium was set for first class tobacco of the Oriental variety, and the lowest for the fifth class of the Burley variety. To be eligible for a premium, producers had to comply with planting density. In 2006 the premiums were temporarily changed to area payments and since 2007 tobacco production subsidies have regained the form of payments for quantity and quality delivered.

Dairy premium

The dairy premium has the longest (decades-old) tradition in Serbian agricultural policy. Payment to milk producers since 2001 is made by the dairies, who forward government money to farmers. The government payment is made monthly to the account of dairies that send lists of dairy producers with the quantities of milk supplied. The premium is defined by quantity of milk delivered, and depends only on the region in which the milk is produced (higher premiums for producers in the hilly-mountainous regions compared to those in the plain regions). The system sought to encourage farmers to sell their milk to dairies rather than through the informal market but had no wider objectives in terms of milk quality or structural change. From 2004-2008, the dairy premium was permanently reduced.

The implementation of dairy premium in 2009 and 2010 differs significantly from the system in the previous period. In 2009 the milk premium was offered only to family holdings, not legal entities, although in 2010 an exception was introduced so that legal entities with at least 90 per cent state capital can receive subsidy (in practice meaning just the 'PKB' farm which is under the jurisdiction of the Belgrade City Assembly). Also, the milk premium scheme in 2009 and 2010 supports producers with higher milk production: in 2009 producers who delivered a minimum of 2,500 litres of milk every quarter (typically requiring at least 4 cows) were eligible for a premium of 1 RSD/litre of milk in compliance with the 'Regulation on quality of milk'. The Regulation stated that the premium should increase by 1 dinar for each litre of extra class (i.e. EU standard) milk delivered to dairy plants, but in practice this was never implemented. In 2010, only natural persons who are holders of a commercial family farms, or legal persons with at least 90 per cent of the state capital established, are entitled to the premium. The payment is made quarterly for a minimum of 4,000 litres (requiring at least 5-6 cows), up to a maximum of 3,000,000 litres per quarter (or at least 2,000 cows) at a rate of 1.5 dinars/litre. With such a payment scheme small producers are excluded from the premium system. The subsidy regulation issued in 2010 does not include any additional incentive for delivering extra class milk.

Producers of sheep and goat milk can also receive premium for milk delivered to dairies, but the number of farms and quantities of milk involved are extremely small.

Direct payments to producers

In 2006 a budgetary support system was established through direct payments per hectare of sown area and per head of livestock. Eligible for these payments were farms included in the Register of agricultural holdings, which are in operation and regularly pay their liabilities from short-term and long-term loans.

As of 2007, along with the abolishment of crop premiums, flat rate subsidies on inputs for farms with less than 10 ha, and subsidies for the purchase of fuel, mineral fertilizers and seeds for farms of 11-100 ha, have become a basic instrument of direct producer support (Table 7.3). The basis for this type of payment for farms over 10 ha are receipts for purchased fuel and plant protection products. Practically, this kind of support in Serbia represents subsidies of inputs, combined with a flat rate amount. In recent years, input subsidies have become more important. Previously, input subsidies focused on breeding livestock, diesel fuel and interest rates on short-term loans. Recently, however, support was extended to calves for fattening, mineral fertilizers, seeds, and insurance premiums.

Since 2009, an additional requirement for the entitlement to the use of financial incentives is that the holder of the agricultural farm is insured at the Pension Insurance Fund, and regularly settles its obligations towards the Fund. Direct payments are implemented for a surface that has been reported in the Register of Agricultural Holdings, and payment is made for an area of 0.5 to 100 ha through submitted invoices for purchased raw materials (seeds, fertilizer and diesel fuel), which serve as proof of performance of manufacturing.

Table 7.3. Amount for reimbursement of raw materials by crops (euro/ha).

	2006	2007	2008	2009	2010
Cereals, industrial crops, vegetable, fodder crops	0	100	123	128	136
1.1.Oilseed rape	83	100	123	128	136
1.2.Oil pumpkin	83	100	123	128	136
1.3.Tobacco	773	0	0	128+212	136+194
1.4.Sugar beet	0	100	123+50	128	136
1.5.Hop	238	100	123	128	136
Flowers	0	0	0	128	136

Source: Own calculation based on GoRS (2011a).

Table 7.4. Number of beneficiaries and disbursed funds for input subsidies/reimbursement.

Year	2006	2007	2008	2009	2010
Number of beneficiaries	73,231	197,411	270,678	84,218	N/A
Total disbursed (million euro)	22.3	90.3	175.3	78.9	22,3

Source: GoRS (2011a).

Table 7.4 above shows that the number of farmers getting the subsidy for their input purchases differs from year to year. In 2009 only 84,000, which is about 12 per cent of farmers, got these direct payments. This means that only a very limited number of farmers were eligible for these subsidies.

Reimbursement for breeding livestock

The objective of this measure is the production and supply of quality breeding of livestock (cows, sheep, goats, sows) and creating conditions for increasing the scope and improving the quality of milk and meat production. A criterion was set so that legal and natural persons and entrepreneurs registered in the Register of Agricultural Holdings are entitled to incentives for bred or purchased heads of livestock which remain on the farm and are used for further reproduction in a period of 4 years for cows, goats and sheep and of 3 years for pigs. Reimbursement is made once during reproduction. The system also applies to bee hives and payment is made for the selected queen bees once during the year for the formation of new swarms.

Stimulation for cattle fattening

Subsidies for fattening young bulls were introduced in 2007 and promoted as the support initiative '90 euro per bull'. Objective is the increasing of the beef meat production and meeting the needs of the domestic market for beef meat, and increased exports of beef to foreign markets. This was the first measure of direct support that was granted in the form of payment per animal. The regulation anticipated subsidizing the raising of fattening bullocks (Spotted Domestic cattle of the Simmental type, Simmental and other breeds). Incentives were granted if the animal had been produced in the farmers' own breeding stock or had been bought to be fattened on the recipient's farm for at least 180 days and was intended for meat production.

Genetic improvement of dairy cattle - subsidy for quality breeding heifers

In 2008, a subsidy for genetic improvement of dairy cattle production has also been introduced. In public it has become known as the '70 euros per registered cow' incentive. The aim of this measure was to increase the number of head of livestock under the control of selection services, upgrading the breeding composition of dairy cattle, increasing milk production per head of cattle, and satisfying the needs of the domestic market for milk and placement of surplus products to the markets of neighbouring countries. The payment scheme implemented through disbursement of an annual amount as direct payments per head of livestock. The criteria was set so that legal and natural persons may apply for obtaining the payments, provided that they are enrolled in the Register of Agricultural Holdings, and that heads of livestock which qualify for payments are used for milk production on their own farms.

Stimulation for genetic improvement of livestock breeding

The incentive has been introduced in 2009. In 2010 payments for high quality cows for herds of at least 5 cows (4 cows in LFAs) and up to a maximum of 100 cows are 125 euro per heifer, and 30 euro/head for sheep and goats for a flock size of 30-150 sheep or 10-150 goats (slightly higher than in 2009). The aim of the measure was to encourage production of quality breeding of livestock and to increase the degree of cattle, swine, sheep and goat breeding, which provide conditions for increasing efficiency, range and quality of milk and meat production.

Beneficiaries of this support are the owners of agricultural holdings enrolled in the Register of Agricultural Holdings, provided that they have obtained the status of insured farmers in the Republic Fund for Pension Insurance, and paid the principal debt for contributions to pension and disability insurance for 2008 and 2009. Right to the use of incentives is also granted to beneficiaries of pension acquired by the agricultural production, if they are holders of commercial farms.

Impact evaluation of market and direct producer support measures

The current situation in the livestock sector is the result of the combined influence of two historical factors: one is the unstable support to producers, which varied in scope, purpose and beneficiaries, and another is the unstructured market chains, which is partly a consequence of the existing grey economy.

The fact that poor policies resulted in a livestock decline is argued by the fact that the basic characteristic of the support, provided to this sector as of 2004 onwards, was its inconsistency with the horizontal and vertical relations within the sector. Namely, while the milk premiums were permanently and significantly reduced (from 45 million euro in 2004 to around 5 million euro in 2010), a number of measures which supported the production of beef increased. These initiatives for the support of fattened cattle have been quite futile, since the number of calves decreased due to the decline in milk producers' interest for this production. In addition, the preconditions for the use of incentives have changed from year to year and became more restrictive towards small producers, who have practically provided the biggest contribution to production. According to the 2002 Census, the farms with area of up to 5 ha have about 45 per cent of cattle and sheep, 55 per cent of pigs and 65 per cent of poultry. Specifically, as for the production of milk, the producers who delivered less than 2,800 litres per year were exempted from support in 2009, and that limit was even raised in 2010 to as much as 4,000 litres per year. This means that practically all farmers who had less than 4-6 cows were not eligible for subsidy! Many of these farms withdrew from milk production and the number of cattle significantly decreased. In order to overcome the effects of the decline in the number of cows, the import of high quality breeds was initiated. However, when the milk premium decreased below an acceptable limit, slaughter and sale of expensive, imported heads of cattle was conducted at slaughter cattle price. Numerous support measures were irrational and reflected complete disorientation. On the one hand, measures for improving the situation caused by previous bad decisions were introduced, and on the other hand there were elements of overlapping support aimed at certain beneficiary groups (see Table 7.4).

The milk premium (RSD 5 per litre) and a RSD 25 thousand subsidy per head of cattle remained in force within the programme of support adopted in February 2011. Although significant, these funds will be practically allocated for correction of (big) mistakes of the previous funding system, which brought this sector to an unenviable position.

Structural and rural development measures

From 2004-2009, the share of rural development support in the overall budget was low and less than objective needs. In the 2004-2005 programme there was only one measure related to rural development which integrated several very different measures, from direct support to producers (purchase of cattle) to support for investments in agriculture and support for the certification of organic agriculture, to more general support for the rural population. The measure was intended only for certain users (e.g. regarding support for investments only farmers aged up to 40 or 55, depending on the region, could apply). In the 2007 programme, this measure was divided into four measures, and in the 2008 programme into five separate measures that have been rounded according to the objective and expanded in terms of users (no age restrictions, different shares of public funds depending on the area).

The 2008 programme is much closer to EU programmes for rural development, in terms of content, users and methods of implementation. Since 2008 this type of support generally follows the distribution of the second pillar of CAP according to the axis, wherein the support for Axis I measures is the highest. This approach is similar to that which is applied to the EU-10 countries that have used the SAPARD funds.

During the observed period 2006-2008 most of the funds were used for 'increasing the competitiveness of farms' and restructuring agriculture by supporting investments for machinery and equipment (Figure 7.3). The largest part of this support was related to subsidizing interest rates on loans intended to renovate machinery. Furthermore, in 2006, large amounts (over 40 million euro, or 50 per cent of support for rural development) were spent for support to non-commercial farms, which was distributed as linear payments of approximately 490 euro per member of a registered farm over the age of 55. The farm households, whose members received such a grant, were not entitled to other types of incentives from the agricultural budget (except milk premium). This measure was supposed to provide a stable income of elderly households, to ensure support to their living standard, and to contribute to the vitalization of land markets.

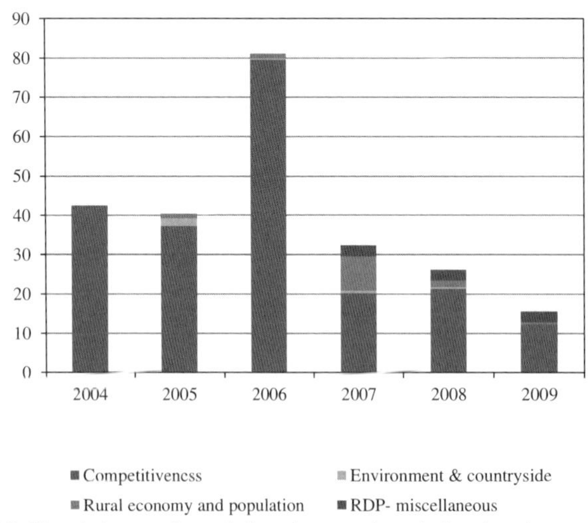

Figure 7.3. Breakdown of rural development and structural support measures (million euro).
Source: Own calculation based on MAFWM (see footnote Table 7.1).

From 2007-2008, funding for subsidizing interest rates was significantly reduced, and support for the revitalization of the land market, for producer's groups and associations, and for the improvement of soil quality were abolished. Apart from subsidizing interest rates on loans and support for non-commercial farms, the following activities have also been supported: improvement of fruit and vegetable production by subsidizing parent plantations; revitalization and eradication of neglected and infected plantations; production of wine and other products with protected designation of origin; genetic improvement of dairy cattle and milk quality.

Among the measures intended for 'environmental protection and the countryside', those for the preservation of plant and animal genetic resources, the development of organic production, protection from erosion and the management of regional waterways have all been implemented. Support to this group of measures has, however, been very modest, amounting to some 4 per cent of total expenditures related to rural development in 2005 only.

Quality of life in rural areas and diversification of the rural economy has been supported modestly, bearing in mind the problems of rural poverty, the transitional surplus labour force, orientation to the development of multifunctional agriculture and the diversification of the rural economy. Since 2005, this type of support has ranged approximately 1 million euro annually, excluding 2007, when the investments were significantly higher (about 8 million euro, item 'Rural economy and population in Figure 7.3). The reason for such a high increase of funds in 2007 is that the National Investment Plan projects for the reconstruction of rural infrastructure were registered in the sub-account of the Ministry of Agriculture. Projects for the promotion of rural tourism as a form of diversification of the rural economy have been financed since 2007 through the refurbishment of facilities and education (maximum 12,500 euro per project).

Finally, since 2007 the strengthening of local partnerships and the capacity of local rural stakeholders has been supported by LEADER-like measures; the founding of rural information centres has been co-financed, as have the establishment of rural networks and the education of local action groups. It should be stressed that apart from the budget incentives to rural development presented, other funds, without precise comparative records, have also been used.

This primarily refers to funds for the reconstruction of rural infrastructure, which were distributed through several ministries. Also, significant funds were invested in support for the renewal of mechanization, land amelioration, the introduction of food quality standards and the promotion of local products and events in the region of Vojvodina. These activities were financed from the budget of the Autonomous Province of Vojvodina, and the programs were executed through public tenders. Rural development was also financed from local/municipal budgets. Based on incomplete data, it has been estimated that the amount of these budgetary incentives has been increasing, as has the number of municipalities having specific budget lines for agriculture and/or rural development. The most supported activities are local events, participation in trade fairs, the promotion of local products and values, tourism potential and the like. In addition, most municipalities that have agricultural budgets also allocate funds for the artificial insemination of cattle, plant protection, prognostic services, etc., while some municipalities also co-finance agricultural loans.

Recognizing the heterogeneous socio-economic characteristics of rural Serbia, measures to support rural development have been formulated so that users in less favoured areas receive more funds primarily because of less favourable natural conditions for agricultural production, but the same applies for regions with protected natural assets or national parks. These benefits are reflected in higher rates of return on the invested funds compared to other regions (as was previously mentioned for higher milk premiums). Less favoured areas according to EU standards have not been defined, but some efforts are being made to establish as objective criteria as possible for their selection. The 2008 program also envisaged support for the members of vulnerable social groups in the rural population (Roma, refugees, displaced persons, socially disadvantaged individuals and returnees from urban areas). Due to a lack of funds, this regulation was not implemented in full as had been planned. On the other hand, what had a great impact on the economic position of rural households was a reduction in support to farms whose owners were over 65 years old, as well as the growth of agricultural products and input price disparities in 2009.

General measures related to agriculture

The need for general support measures is significant, bearing in mind the institutional disorder of the sector and the technical and technological unpreparedness of institutions to meet complex requirements associated with agricultural policy reform. Generally, funds for financing various departments are being increased, but it should be emphasized that these funds are relatively small, indicating weak development levels, primarily of general services for agriculture (improving cadastral records, extension service, etc.). Within the general services sector, the greatest proportion of support funds was directed to extension services or to financing the agricultural extension service (34-56 per cent). Activities performed by the extension service (quality control, recommendations and advice to manufacturers, etc.) are financed from the Agrarian budget, based on a specific contract.

In recent years, general services in agriculture also included support for hydrometeorological stations, for monitoring weather conditions, natural disaster warning through the media, anti-hail interventions, forecasts for the occurrence of plant diseases and pests. A smaller portion of funding was used for the needs of veterinary inspection, co-financing soil quality control and the work of the extension bodies of the Ministry. This part of the support is certainly greater than the finances recorded, as one part has been executed through other ministries (e.g. the Ministry of Science is financing research projects related to technical and technological innovation and raising standards of food safety and quality, whose results will be applied in practice) or other directorates of MAFWM. Since 2001, a significant part of the activities in this group of measures has been financed by donor or development project funds (Bogdanov and Bozic, 2010).

7.5 Concluding remarks

Reforming agricultural policy in Serbia began after the political changes of 2000. The first steps were oriented to the establishment of more efficient inspection services and combating the informal economy. Another priority has been to increase the volume of production, which after a decade of isolation was at a very low level. The main instruments of support were based on production (premiums and direct payments related to yields) as well as reimbursements for the inputs used. Thanks to these incentives within a short period agricultural production increased significantly and resulted in surpluses and increased exports. The main instruments of market policy were high customs duties for sensitive products, for which it is estimated that there is a possibility to enhance the competitiveness before liberalization, and export subsidies. Market intervention purchases and mechanisms of price policy were not used.

In 2005, the government adopted the Agriculture Development Strategy, which was replaced in 2010 by Agriculture Development Strategy for 2010-2020 that will be implemented through the National Agricultural Programme of the Republic of Serbia for 2010-2013 and the National Rural Development Programme of the Republic of Serbia for 2011-2013.

Quantitative analysis of the Agricultural budget clearly indicates a significant variation over time. There was no stability of the total budget for agricultural support measures, which is a sign that policy is still in transition and that no long-term framework exists. Financial resources are not specified for several years in advance, and they change depending on the annual state budget and the share of agriculture it contains. The lack of continuous and consistent policy, i.e. clearly-defined preferences and the development of mechanisms for their implementation, cause inefficiency and had the outcome that the overall results of the policy are likely to be below the expected and objectively possible levels.

Although the dynamic changes in volume of support for agricultural producers were painful, perhaps even more problematic were frequently radical changes in the type and form of payment. A system of registration of holdings was introduced as of 2004, which gradually developed as registration formalities became a prerequisite for access to public support. Terms of eligibility for support are becoming more complex, and the selection criteria were more biased towards the large producers. An unpredictable support system gives producers a sense of instability and uncertainty, which discourages farmers to invest. Half of the total number of farms are registered as potential beneficiaries of budget support. On the other hand, only 10-15 per cent of farm households benefit from direct support per hectare, and only for arable land.

Contrary to EU practice, a prerequisite for this support is the receipt of the purchase of inputs instead of meeting standards of quality, environment and natural resources, animal and plant welfare. Market support measures and direct support to producers have been particularly inconsistent and without clear goals with regard to livestock production. Support measures have been radically changed during last few years, which lead to counterproductive effects or to an overlap of support for the same purpose.

Support for rural development has become more actively implemented since 2004. Most of the funds have been used (as it was the case in the NMS) to support improving competitiveness of farms, and the basic mechanism has been subsidies of interest rates on loans for farm mechanization and equipment. There are some elements of support based on the model of the LFA directive, but without clearly defined rules. In recent years there have been some attempts to deal with some sensitive social issues through support for rural development, but this practice was later abandoned (support for Roma population in rural areas, refugees, non-commercial farms, etc.).

Serbian agricultural policy requires fundamental reforms at all levels. The model of support to agriculture that is applied in Serbia is significantly different from the EU practice and will have to adjust. Adjustment will require a new (accredited) institutions, greater transparency and access to support for a wider range of users.

Literature and websites

Bogdanov, N., *Mapping strategies, institutions and funds associated with rural development in Republic of Serbia*. UNDP, internal documentation, UNDP, 2008.

Bogdanov, N. and D. Bozic, 'Review of agriculture and agricultural policy in Serbia.' In: *Agriculture in the Western Balkan Countries*. T. Volk (ed.), IAMO, Halle, Studies on the agricultural and food sector in Central and Eastern Europe, Vol. 57, 2010.

Bogdanov, N., T. Volk, M. Rednak and E. Erjavec, *Analysis of direct budgetary support to agriculture and rural development in Serbia*. PRSP, Belgrade, 2010. Available at: www.prsp.gov.rs/download/Analizadirektnepodrskepoljoprivrediruralnomrazvoju.pdf.

GoRS, *Answers to the European Commission's questionnaire*. 2011a.

GoRS, *First national report on social inclusion and poverty reduction in the Republic of Serbia.* Belgrade, 2011b.

Jongeneel, R. and H. Brand, 'Direct income support and cross-compliance in EU policy for agriculture, food and rural areas.' In: *EU policy for agriculture, food and rural areas.* Oskam, A., G. Meester and H. Silvis, (eds.), Wageningen, Wageningen Academic Publishers, Wageningen, 2011.

Massot Marti, A., *The First Pillar of the CAP: I. The Single Common Market Organization. European Parliament.* 2010. Available at: www.europarl.europa.eu/parliament/expert/FTU_4.2.3.html.

Milanovic, M., 'The research of the Serbian agrarian development, position in Europe.' In: *Megatrend Review* 3 (2006) 2.

Silvis, H. and R. Lapperre, 'Market, price and quota policy: from price support to safety net.' In: *EU policy for agriculture, food and rural areas.* Oskam, A., G. Meester and H. Silvis (eds.), Wageningen, Wageningen Academic Publishers, 2011.

Chapter 8

8 Serbia's foreign trade position

8.1 Introduction

8.1.1 An overview of Serbia's major trade relations

Total merchandise trade and balance of payments

Foreign trade has been growing in importance for the Serbian economy since 2005. In that year, Serbia's total exports and imports of goods were 18.0 per cent and 41.9 per cent of GDP, respectively. These shares grew steadily in the period up to 2008, with a small decline in the export share and a rather significant fall in the import share in 2009. Trade shares increased again in 2010, and Serbia's deficit on merchandise trade amounted to almost 20 per cent of the country's 2010 GDP (26.5 billion euro).

Table 8.1. Serbia's merchandise trade flows, billion euro, 2004-2010.

	2004	2005	2006	2007	2008	2009	2010
Exports	2.98	3.67	5.10	6.43	7.43	5.96	7.39
As % of GDP	15.6	18.0	22.0	22.4	22.5	22.0	27.8
Imports	8.94	8.53	10.46	13.36	15.58	11.16	12.62
As % of GDP	47.0	41.9	45.1	46.6	47.1	37.4	47.5

Source: Statistical Office of the Republic of Serbia.

Table 8.2 shows the main destinations and sources of Serbia's merchandise trade flows. The EU is by far both the largest exporter to, and importer from Serbia. Italy and Germany are the most important EU trading partners of Serbia. The trade figures also indicate the increased orientation of Serbia towards the European Union since 2005. In part this is the effect of the EU membership of neighbouring countries like Hungary (in 2004) and of Rumania and Bulgaria in 2007. Serbia's exports to these three countries amount to around 7 per cent of total exports, while Serbia's imports from these countries are 8 to 9 per cent.

Next to the EU, Serbia's main export markets are the ex-Yugoslavian republics, to which Serbia's export share increased from 30 per cent in 2004 to 35 per cent in 2009, but declined to 32 per cent in 2010. Bosnia/Herzegovina and Montenegro are the main export markets, whereby since 2008 the export share to Bosnia/Herzegovina is higher than the most important EU-market (Germany). On the import side, these countries are much less important to Serbia, as they account for 10-11 per cent of Serbia's imports. These figures show that Serbia has significant positive trade balance with the former Yugoslavian republics -- around 2 billion euro in 2010. It also indicates that the country has a strong interest in maintaining access to its neighbours' markets.

The Russian Federation is traditionally an important trading partner of Serbia, as the country imports a significant part of its energy commodities from Russia, largely oil and natural gas. Moreover, Serbia and the Russian Federation signed a Free Trade Agreement, that encourages bilateral trade (see section 8.3).

In recent years China has gained importance as a source for Serbia's imports.

Table 8.2. Direction of trade in goods, 2005-2010.

Trading partners	**2005**	**2006**	**2007**	**2008**	**2009**	**2010**
		Exports by destination, % share				
EU	32.1	55.6	57.2	54.1	53.9	57.1
of which Italy	14.3	14.4	12.4	10.3	9.8	11.4
of which Germany	9.6	9.9	10.6	10.4	10.4	10.9
Russian Federation	4.9	4.8	5.1	5.0	4.2	6.5
Rep. Bosnia & Herzegovina	16.3	11.6	11.8	12.2	12.2	11.1
Slovenia	4.1	3.9	4.6	4.6	4.1	4.3
Croatia	4.3	3.9	3.7	4.0	3.3	3.1
Macedonia	5.7	2.4	5.0	4.5	5.1	4.9
Montenegro	.	9.6	10.8	11.7	10.0	8.2
Others	32.5	8.2	1.8	3.9	7.2	4.8
Total	100	100	100	100	100	100
		Imports by origin, % share				
EU	30.6	55.4	60.3	58.2	59.2	57.3
of which Italy	8.6	8.3	9.8	9.5	9.8	8.5
of which Germany	10.3	9.5	12.0	11.7	12.5	10.6
Russian Federation	15.8	16.3	14.3	15.2	12.6	12.9
Rep. Bosnia & Herzegovina	2.8	2.6	2.8	2.8	2.8	3.3
Slovenia	2.7	2.3	3.8	2.7	2.9	3.0
Croatia	2.4	2.5	2.9	2.4	2.5	2.6
Macedonia	1.6	1.5	1.7	1.6	1.5	1.6
Montenegro	0.0	1.1	0.7	0.9	0.0	0.9
China	4.8	5.9	7.4	7.8	7.7	7.2
Others	39.3	12.3	6.1	8.4	10.7	11.2
Total	100	100	100	100	100	100

Source: Statistical Office of the Republic of Serbia. Note that 2005 figures show Serbia and Montenegro together while the data for 2006 and onwards reflects external trade of Serbia.

Serbia's main export items (according to SITC) in 2009 and 2010 were iron and steel, clothes (only in 2009), electronic machines and apparatus (in 2010), cereals and cereal products, fruit and vegetables and non-ferrous metals. These sections accounted for 30 to 35 per cent of the overall exports. The major categories with the greatest import share were oil and oil derivatives, natural gas, electrical machines and apparatus, and industrial machines for general use, which all together accounted for 25-30 per cent of the overall imports.

Trade in agricultural products[1]

In 2010 total agricultural exports amounted to 1.7 billion euro, whereas total agricultural imports was 830 million euro (see Table 8.3). This makes Serbia a net exporter of agricultural products, a position the country had already for several years. Both agricultural exports and imports show a steady increase over the last years, although in 2009 and 2010 agricultural imports were significantly less than they were in 2007 and 2008.

Table 8.3. Agricultural products as a percentage of total merchandise trade.

	2005	2006	2007	2008	2009	2010
Total exports of goods, billion euro	3.67	5.10	6.43	7.43	5.96	7.39
Agricultural products, million euro	790	1,067	1,303	1,398	1,442	1,700
Agricultural products % of total exports of goods	21.5	20.9	20.2	18.8	24.1	23.0
Total imports of goods, billion euro	8.53	10.46	13.36	15.58	11.16	12.62
Agricultural products, million euro	703	846	976	1,163	823	830
Agricultural products % of total imports of goods	8.2	8.0	7.3	10.4	7.4	6.6

Source: Statistical Office of Serbia for total exports and imports; UNCOMTRADE for agricultural exports and imports.

Table 8.4 gives a breakdown of the Serbia's total agricultural exports and imports since 2005. In 2005, the most important export products were sugar (HS17), fruits and nuts (HS08), cereals (HS10) and beverages (HS22), totalling 53 per cent of all agricultural exports. In 2009 these products accounted for 50 per cent, the result of declining exports of sugar and increased exports of beverages compared to the 2005 levels. Agricultural imports are more diversified. The four major imported products (fruits and nuts, coffee/tea, miscellaneous edible preparations and beverages) make up about 30 per cent of Serbia's agricultural imports. There is relatively little trade in meat and dairy in either direction.

[1] Definition of agricultural products: Chapters 01-24 of the Combined Nomenclature (excluding fish 03 and fish/seafood from other chapters) is the usual definition adopted by the European Community for its trade statistics. Note that additional products are included in the definition of agricultural products used with the WTO's Uruguay Round Agreement on Agriculture (see par. 8.3). The definition the FAO uses in FAOSTAT includes some of the additional WTO products but omits a few of the more processed items that are included in the regular EU definition.

Table 8.4. Serbia's agricultural trade values: selected product categories, 2005-2009

Product category					
Total agricultural products	**Value of Exports (Eur mn)**				
	2005	**2006**	**2007**	**2008**	**2009**
Total agricultural products	790.3	1,067,1	1,303,9	1,397,9	1,441,5
Meat	11.0	32,6	41,5	30,1	21,6
Dairy and eggs	7.7	28,4	33,4	40,6	40,7
Vegetables, fruits (1)	208.8	256,6	335,9	319,4	320,6
Coffee, tea, maté and spices	5.7	7,0	10,4	9,2	10,0
Cereals (2)	139.7	209,4	189,5	178,8	311,0
Products of the milling industry	11.2	27,8	65,6	76,7	40,5
Oilseeds	16.6	18.9	20,4	28,1	26,1
Animal or vegetable fats and oils	36.3	29.0	76,4	105,3	93,6
Preparations of meat, fish, etc.	17.3	32,3	31,8	35,9	29,1
Sugars	141.1	133,4	120,5	118,1	111,4
Cocoa and cocoa preparations	30.0	39,7	42,9	47,9	49,4
Miscellaneous edible preparations	29.6	50,2	51,1	55,9	55,5
Beverages	44.8	84,6	112,3	141,3	140,3
Animal fodder	14.8	21,2	43,7	48,0	44,1
Tobacco	3.0	10,8	21,2	30,7	40,2
	Value of Imports (Eur mn)				
	2005	**2006**	**2007**	**2008**	**2009**
Total agricultural products	703,5	845,7	976,2	1,162,7	823,5
Meat	5,7	5,0	7,1	17,0	15,3
Dairy and eggs	8,6	8,0	16,0	25,6	14,8
Vegetables, fruits 1)	126,8	150,4	189,6	224,4	173,0
Coffee, tea, maté and spices	48,7	50,4	60,7	75,1	53,3
Cereals 2)	32,5	42,6	53,5	65,8	47,7
Products of the milling industry	9,1	9,3	8,7	9,9	4,5
Oilseeds	22,1	30,2	37,3	55,7	44,2
Animal or vegetable fats and oils	15,6	32,8	33,7	44,3	37,4
Preparations of meat, fish, etc.	21,6	25,0	36,1	45,6	40,1
Sugars	32,6	29,6	32,4	34,1	23,3
Cocoa and cocoa preparations	43,8	48,6	57,7	67,6	42,4
Miscellaneous edible preparations	69,8	61,3	65,9	81,0	49,5
Beverages	24,6	50,0	58,5	62,2	46,1
Animal fodder	39,3	45,4	37,0	49,5	27,8
Tobacco	68,6	79,3	59,5	60,2	38,7

Notes: (1) Includes edible vegetables, edible fruits and preparations of vegetables and fruits HS 07,08 and 20); (2) includes cereals preparations, etc. (HS 19). Note that Agricultural products are defined as Chapter 01-24 of the CN.
Source: UN Comtrade.

The ex-Yugoslavian republics are important trade partners of Serbia. This also holds for agricultural products. Almost half of Serbia's agricultural exports has a destination in one of the former republics (see Table 8.5; Slovenia is left out, as it is part of the EU). About a quarter of Serbia's total agricultural imports are from these countries. Trade in agricultural products between Serbia and its neighbouring countries is rather diverse. One of the major commodities Serbia is exporting in the region is cereals, or cereal products/preparations. Beverages are a major item in both export to and import from the ex-Yugoslavian republics. Serbia is a (traditional) net exporter of agricultural products to each of these countries, except for Croatia.

Table 8.5. Serbia's agricultural trade with (non-EU) ex-Yugoslavian republics (2008 data, values in million euro).

Partner country	Export value	Major export commodity	Import value	Major import commodity
Bosnia & Herzegovina	287	Cereals, beverages	94	Animal fodder, sugar & sugar confectionery
Montenegro	251	Live animals, meat, dairy, beverages	28	Beverages
Macedonia	111	Products of cereals	74	Vegetables, beverages
Croatia	47		73	Oilseeds, tobacco, fruits

Note: Agricultural products defined as Chapter 01-24 of the CN.
Source: UN Comtrade.

Serbia's agricultural trade with the European Union

Table 8.6 below shows the agricultural trade flows between Serbia and the EU in 2008, 2009 and 2010. Serbia had a positive trade balance on agricultural trade with the EU, amounting to 45 million euro in 2008, 254 million euro in 2009 and 247 million euro in 2010. Serbia's main (five) agricultural imports from the EU are miscellaneous edible preparations (HS21), fruit (HS08), beverages (HS22), residues and waste from the food industries (HS23) and tobacco (HS24), accounting for about 40 per cent of total agricultural imports from the EU in 2010. Serbia's main agricultural exports to the EU are fruit (HS08 -- especially raspberries, sour cherries and black berries), cereals (HS10), sugar (HS17), animal or vegetables fats and oils (HS15, of which soybean oil and sunflower seed oil are Serbia's major export products to the EU). These five product categories account for 73 per cent of Serbia's total agricultural exports to the EU in 2010. Figures indicate that Serbia's agricultural export to the EU continues to increase, while EU exports to Serbia did not reach the 2008 level in 2009 and 2010.

Table 8.6. Bilateral agricultural trade EU - Serbia, in million euro.

Product category (2 digit HS-code)	EU exports to Serbia			EU imports from Serbia		
	2008	2009	2010	2008	2009	2010
01 - Live animals	8.6	8.4	5.2	.2	0.2	0.2
02 - Meat and edible meat offal	11.0	14.2	11.6	7.9	4.5	4.8
04 - Dairy produce; birds' eggs; natural honey	20.9	10.3	23.4	0.8	1.2	4.2
05 - Other products of animal origin	2.0	1.5	1.7	1.0	0.6	1.2
06 - Live plants and floricultural products	16.8	12.5	14.0	0.8	0.9	1.1
07 - Edible vegetables, plants, roots and tubers	18.8	11.8	13.5	29.5	34.6	36.6
08 - Edible fruit and nuts; peel of citrus fruits or melons	38.0	40.5	39.0	182.7	186.0	189.7
09 - Coffee, tea, maté and spices	6.9	6.	7.1	5.0	6.1	7.1
10 – Cereals	28.4	8.0	9.2	30.8	179.3	112.0
11 - Products of the milling industry; malt; starches	4.9	3.5	3.7	6.8	4.5	4.1
12 - Oil seeds and oleaginous fruits	25.3	21.6	29.8	9.5	11.2	15.5
13 - Lacs; gums, resins and other vegetable saps and extracts	3.8	4.0	3.8	0.0	0.0	0.0
14 - Vegetable plaiting materials	0.2	0.1	0.1	1.0	1.0	0.8
15 - Animal or vegetable fats and oils	35.4	11.9	21.5	56.5	40.8	73.1

Table 8.6. Bilateral agricultural trade EU - Serbia, in million euro.

Product category (2 digit HS-code)	**EU exports to Serbia**			**EU imports from Serbia**		
	2008	**2009**	**2010**	**2008**	**2009**	**2010**
16 - Preparations of meat, fish	18.3	17.6	17.5	0.1	0.1	0.1
17 - Sugars and sugar confectionery	12.4	11.0	10.7	113.0	95.6	104.1
18 - Cocoa and cocoa preparations	41.1	36.8	36.4	3.1	2.9	3.3
19 - Preparations of cereals, flour, starch	21.3	17.5	20.5	5.8	7.8	9.4
20 - Preparations of vegetables, fruit, or nuts	18.1	17.4	18.9	19.6	17.8	30.8
21 - Miscellaneous edible preparations	56.7	50.7	55.4	10.0	12.5	14.6
22 - Beverages, spirits and vinegar	24.5	27.5	15.1	8.4	13.5	13.8
23 - Residues and waste from the food industries	25.2	25.2	28.6	13.3	18.1	23.2
24 - Tobacco and manufactured tobacco substitutes	25.0	31.0	26.2	3.2	4.3	9.6
Total	463.6	389.2	412.9	508.7	643.5	659.3

Note: Chapter 1-24 of the Combined Nomenclature (excluding fish [03] and fish/seafood from other chapters).
Source: Eurostat.

8.2 Trade policy

Institutional aspects

The ministry of Economy and Regional Development is responsible for Serbia's trade policies, in consultation with other ministries and external bodies. The umbrella regulation is the Law on Foreign Trade Transactions ('Official Gazette of RS', No. 36/09), which was adopted in May 2009. This Law, which is in compliance with the rules of the World Trade Organization (WTO), prescribes the basic elements and rules of the foreign trade policy under which it is possible to introduce measures affecting foreign trade transactions. This legislation deals, inter alia, with the export/import of goodsfavoured nation status, quantitative limitations and permits, safeguards, countervailing and antidumping duty issues, temporary regimes and related measures. The Law is based on principles of competition and transparency, but also allows state intervention in cases that may harm the economy. It identifies several situations in which the state can investigate and respond to illegitimate and unfair competitive threats from a foreign state or business enterprise, including instances in which a foreign state provides subsidies, thus giving the entity a privileged position against domestic companies.[1]

The implementation of the trade measures for agricultural products is carried out by the Ministry of Agriculture, Trade, Forestry and Water Management (MATFWM). This Ministry is responsible for the veterinary and phytosanitary services at the border. The Ministry lays down and checks veterinary and sanitary conditions for import, transit and export of consignments of animals, food, feed and feed components, products of animal origin, animal by-products and reproductive material. Also the MATFWM monitors the

[1] For a full text see GoRS, 2011, annex 30 to Chapter 29 Customs Union.

epizootic situation in the exporting country and the transit country, as well as the equivalence of veterinary and sanitary control systems related to risk analysis in foreign trade.[1]

Serbia is in the process of becoming a member of the WTO. Serbia applied for membership in 2004. There have been several meetings of a working party for the accession of Serbia in which Serbia negotiates market access for goods and services bilaterally with interested WTO members. Serbia has entered in a number of preferential agreements with trading partners. In the winter of 2010/2011 important further steps have been taken that bring WTO membership of the country very close (see also below).

Agricultural trade policy measures

The mainstay of the Serbia's agricultural trade policy is the import tariff. Next to ad valorem tariffs (% of the import value), Serbia also uses specific duties (an amount per tonne). Serbia also applies export subsidies for agricultural products, but only to a very limited extent (see section 7.4, Table 7.2).

In the early 2000s, while still in a union with Montenegro, Serbia implemented substantial trade liberalization reforms, including reduction of tariff and nontariff barriers, elimination of import quotas, reduction of import licensing requirements and prohibitions, and restructuring and simplification of customs procedures. As a result, based on its 7.3 per cent simple average MFN applied tariff, Serbia's trade regime is comparable to that of the average European and Central Asian (ECA) country and more liberal than that of the average upper-middle-income country (6.8 and 9 per cent, respectively; see World Bank, 2010). However, only 3.3 per cent of its tariff lines of agricultural products have zero MFN duties, while 45 per cent have duties higher than 15 per cent (see Table 8.7). Similar to the majority of other countries in its comparator groups, Serbia is more protective of its agricultural goods (14.2 per cent tariff) than of its non-agricultural ones (6.3 per cent tariff).

---

[1] The MATFWM has a major responsibility maintaining food safety – see the Food Safety Law (Official Gazette RS No 41/09) that complies with the WTO/SPS Agreement and EU legislation. As an administrative body within the ministry the Directorate of National Reference Laboratory performs tasks within the field of food safety, animal and plant health, etc.

Table 8.7. Tariffs: Frequency distribution summary and duty ranges.

Frequency distribution		**Duty-free**	**0 <= 5**	**5 <= 10**	**10 <= 15**	**15 <= 25**	**25 <= 50**	**50 <= 100**	**>100**
		Tariff lines and import values (in %)							
Agricultural products									
MFN applied	2009	3.3	34.4	10.6	6.9	24.7	20.1	0	0
Import value	2008	0.6	28.5	12.7	12.3	20.8	25.3	0	0
Non-agricultural products									
MFN applied	2009	1.1	61.5	23.9	4.5	8.6	0.4	0	0
Import value	2008	2.3	65.3	17.6	6.8	7.9	0.1	0	0

Note (a): Frequency distribution by duty ranges. The shares by duty ranges in the frequency distribution are based on the pro rata shares of tariff line level duties in the standard HS six-digit subheadings. For example, if there are two tariff lines in one HS six-digit subheading, one with a duty of 10 and one with a duty of 20, then half of the HS six-digit subheading is allocated to the 5-10 range and the other half to the 15-25 range. The frequency distribution includes AVEs (ad valorem equivalence) whenever those were calculated.
Source: WTO, at www.wto.org, accession countries, Serbia. Data from January 2011.

Table 8.8 below indicates Serbia's tariffs by product groups. Except for beverages and tobacco, where duties are higher, the maximum duties that are applied within each of the product categories shown are 30 per cent. This rate is significant, yet average tariffs applied are generally much less than this maximum rate, although in agriculture the animal products (including dairy), cereals and beverages and tobacco do show an average rate of more than 20 per cent.

Table 8.8. Tariffs by product groups.

Product groups	**MFN applied duties**		
	AVG (a)	**Duty-free (%) (b)**	**Maximum duty rate (%)**
Animal products	22.6	0	30
Dairy products	24.0	0	30
Fruit, vegetables, plants	14.4	0	30
Coffee, tea	10.4	0	30
Cereals & preparations	21.6	0	30
Oilseeds, fats & oils	7.2	0	30
Sugars and confectionery	10.3	0	25
Beverages & tobacco	24.1	0	42
Cotton	0.0	100.0	0
Other agricultural products	3.9	13.4	20

Notes; (a) AVG: Simple average of MFN applied duties; (b) Duty-free in %: Share of duty-free HS six-digit subheadings in the total number of subheadings in the product group. Partially duty-free subheadings are taken into account on a pro rata basis.
Source: WTO, at www.wto.org, accession countries, Serbia. Data from January 2011.

Detailed information (at 10-digit level on applied tariffs (ad valorem, specific and seasonal rates) can be found in the Annex 1 -- Trade Policy, Table 7, to Chapter 11 Agriculture and Rural Development, EC Q&A. A summary is presented in Table 8.9 below at 2-digit level, indicating the range of tariffs that can be found in each product category at the lowest level. The general picture confirms the above indication based on WTO reports that the maximum duties do not exceed 30 per cent. The overview below also provides information about the specific duties, in RSD/kg. These kind of duties are especially applied in case of imports of live animals, meat and dairy products, vegetables, fruit and sugar. Together with the tariffs, these product categories are most significantly protected by import restricting measures.

Table 8.9. MFN applied duties, Serbia.

Product category (2 digit HS code)	Range of applied ad valorem duties (min-max)	Specific tariffs applied to
01 - Live animals	1-30	Live bovine animals, swine, sheep, poultry
02 - Meat and edible meat offal	5-30	Bovine, swine, sheep & goats and poultry, pig & poultry fat, meat offal
04 - Dairy produce; birds' eggs; natural honey	5-30	All products
05 - Other products of animal origin	1-5	
06 - Live plants and floricultural products	10-15	
07 - Edible vegetables, plants, roots and tubers	10-30	All except peas and beans
08 - Edible fruit and nuts; peel of citrus fruits or melons	5-20	Grapes, melons, apples, appricots/cherries etc.
09 - Coffee, tea, maté and spices	3-10	Peppers
10 - Cereals	3-30	
11 - Products of the milling industry; malt; starches	10-30	
12 - Oil seeds and oleaginous fruits	5-20	
13 - Lacs; gums, resins and other vegetable saps and extracts	1-5	
14 - Vegetable planting materials	1-3	
15 - Animal or vegetable fats and oils	1-30	Margarine
16 - Preparations of meat, fish	10-30	Sausages; other prep. meat
17 - Sugars and sugar confectionery	3-25	Cane/beet sugar; other sugar
18 - Cocoa and cocoa preparations	10-30	
19 - Preparations of cereals, flour, starch	15-30	
20 - Preparations of vegetables, fruit or nuts	5-20	
21 - Miscellaneous edible preparations	3-30	Sauces
22 - Beverages, spirits and vinegar	20-30	
23 - Residues and waste from the food industries	1-20	
24 - Tobacco and manufactured tobacco substitutes	10-30	Unmanufactured tobacco

Note: Chapter 1-24 of the Combined Nomenclature (excluding fish (03) and fish/seafood from other chapters).
Source: Annex 1 Trade Policy, table 7, to Chapter 11 EC Q&A.

Serbia in the WTO

Negotiations on membership in the World Trade Organization (WTO) began in 2005. In September 2011 these negotiations were at a rather advanced stage.[1] In October 2010 WTO members reviewed the third revision of the draft Working Party Report containing the reforms to its trade regime undertaken by Serbia in order to comply with WTO rules. Members positively evaluated the progress made in the on-going bilateral negotiations on market access for goods and services, yet also pointed out that some work still remained to be accomplished on some specific issues, notably on trading rights, import licensing, customs, sanitary and phytosanitary issues, intellectual property and services (WTO News, 20 October 2010). Since then Serbia has closed bilateral market access negotiations with several WTO members, also with the EU. On 12 January 2011, Serbia's Minister of Economy Mladjan Dinkic and EU Trade Commissioner Karel De Gucht signed a market liberalization protocol in Brussels as part of Serbia's accession into the WTO. The negotiations for the liberalization were not conducted directly, because Serbia had already signed a Stabilization and Association Agreement and an Interim Trade Agreement with the EU. The protocol is based on those agreements, since they already regulate all relevant trade issues (Balkans.com). The protocol is important in order to complete the negotiations with other WTO members, as the document is expected to encourage other WTO members with whom Serbia is still negotiating.

Trade arrangements

The Central European Free Trade Agreement (CEFTA) is a trade agreement between non-EU countries in Central and South-Eastern Europe. As of 1 May 2007, the parties of the CEFTA agreement are Albania, Bosnia and Herzegovina, Croatia, Macedonia, Moldova, Montenegro, Serbia and UNMIK-Kosovo. Former parties were Bulgaria, Czech Republic, Hungary, Poland, Romania, Slovakia and Slovenia. Their CEFTA membership ended when they joined the EU. When these parties left, it was decided to extend CEFTA to cover the rest of the Balkan states, which already had completed a matrix of bilateral free trade agreements in the framework of the Stability Pact for South Eastern Europe. On 6 April 2006, at the South East Europe Prime Ministers Summit in Bucharest, a joint declaration on the expansion of CEFTA was adopted. The new enlarged agreement was initialled on 9 November 2006 in Brussels and was signed on 19 December 2006 at the South East European Prime Ministers Summit in Bucharest. The agreement went into effect on 26 July 2007 for Albania, Kosovo, Macedonia, Moldova and Montenegro, and on 22 August for Croatia. Bosnia and Herzegovina ratified it on 6 September, while Serbia completed the final legal procedures on 24 September 2007. The agreement aims at establishing a free trade zone in the region by 31 December 2010, but that objective has not been achieved yet.

In February 2011 Albania, Croatia, Macedonia, Montenegro, Moldova and Serbia agreed on more liberalized concessions in trade of agricultural products by signing the Additional Protocol to the Agreement. The Parties abolished all customs duties on imports, all charges having equivalent effect, and all import duties of a fiscal nature in trade between

[1] See WTO news item: Serbia's accession is steadily moving towards an endgame, 30 September 2011 (www.wto.org 2011 news items).

them on all the products other than those subject to bilateral concessions (see Annex 3 to CEFTA 2006 agreement. For Serbia this agreement is a follow-up to earlier agreements in which full liberalization of agricultural trade already had been achieved with Bosnia and Herzegovina, FYR Macedonia and Montenegro (Bogdanov and Bozic, 2010).

The CEFTA links all the Western Balkans and Moldova. It replaces a previous network of more than 30 bilateral FTAs, thereby setting uniform trade rules across the whole of South Eastern Europe. Although not a party to CEFTA, the EU supports the process, which it sees as complementing the Stabilization and Association process.

Serbia signed a Free Trade Agreement with Russia in August 2000 and the list of products, excluded from the Free Trade Agreement, is revised annually. In April 2009, the duty-free regime was extended to the following goods: all drugs, confectionery products, apple juice, malt beer, fresh grape wines, all soaps, wool clothing, refrigerators, freezers and all refrigerating devices, washing and drying machines, wooden upholstered seats, wooden office furniture, sleeping bags, sheets, and similar goods. Outside the Commonwealth of Independent States, Serbia is the only country having a free trade agreement with Russia, which attracts foreign investors to come to Serbia just because of the possibility to export to the Russian market under preferential conditions.

The Free Trade Agreement with Belarus envisages mutual abolishment of customs and non-customs duties in trade between the two countries. There are only a few exceptions to the Agreement, including sugar, alcohol, cigarettes, as well as used cars, buses, and tyres.

The recently signed Free Trade Agreement with Turkey has taken effect on the 1st of September 2010, while the FTA with EFTA members (namely Switzerland and Liechtenstein, since Norway and Iceland are expected to ratify the Agreement at the beginning of 2011) has come to force on the 1st of October 2010.

8.3 Serbia's agricultural trade arrangements with the EU

In 1999, the EU launched the initiative to include the West Balkan countries to which Serbia also belongs, in the process of stabilization and association, with the intention to stabilize the region in a political and economic sense. A stabilization and association process for five countries in Southern Europe, including Serbia, was launched in November 2000, in which assistance and support programmes were developed to contribute to the region's reconstruction and development. By the end of 2005 negotiations for a Stabilization and Association Agreements took-off. It took about 3 years to finalize the talks; on 29 April 2008 the SAA was signed between Serbia, the European Union and its Member States. However, as ratification by Member States for the SAA was delayed (and still is not completed) because of political reasons, an Interim Agreement on trade-related matters was negotiated between Serbia and the European Commission, allowing for the early implementation of trade provisions of the agreement. Currently Serbia's trade relations with the EU are framed by the Interim Agreement on Trade and Trade-related Issues. Serbia decided to implement the provisions of the Interim Trade Agreement (ITA) on 30 January 2009, whereas the EU followed on 1 February 2010.

Agricultural trade between Serbia and the EU is regulated by Articles 24-33 of the SAA, which are Articles 9-18 in the ITA and by the contents of four Annexes and two Protocols. Products include agricultural and fishery products (as listed in Chapters 1-24 of the Combined Nomenclature) and processed agricultural products.

The EU has offered Serbia the immediate abolishment of all quantitative restrictions, customs duties and levies on its imports of agricultural and fishery products from Serbia.

Yet, there are several exceptions. On imports of live bull calves and baby beef, sugar and wine, tariffs are reduced (compared to the MFN tariff level) but not abolished completely, while for baby beef and sugar an annual tariff rate quota applies (of 8,700 tonnes and 180,000 tonnes, respectively) and which are subject to a reduced ad valorem and specific tariffs. Furthermore, of the two product categories vegetables and fruits, the specific tariffs (a fixed amount per tonne) are abolished while the ad valorem tariffs, although reduced, remain. Next, the EU eliminates all customs duties on fish and fishery products originating from Serbia, except for those products listed in Annex IV to which a 70 per cent of MFN duty rate will remain (with a zero rate TRQ for some species).

Protocol 1 of the Interim Trade Agreement describes and lists the processed agricultural products for which duties are set to zero (both for imports into the EU originating from Serbia and vice-versa). Protocol 2 to the Agreement sets forth the regime of mutual trade with wines and alcoholic beverages in terms of preferential quotas, mutual recognition, protection, and control of wine, alcoholic beverages and flavoured wines trademarks. The total quota for preferential export of wine into the EU from Serbia is 63,000 hectolitres. Serbia, on the other hand, allows 25,000 hectolitres of wine from the EU free of customs duties.

Serbia's other concessions to the EU consists of the abolishment of quantitative restrictions and customs tariff from the date of entry into force of the agreement, next to a reduction of tariffs and duties following a step-by-step procedure over a transition period of 6 years. The Annex refers to lists of products that are subject to each of the following regimes:

- Annex IIIa: products for which import duties are reduced immediately at the date of entry of the Agreement into force. The Annex lists 758 tariff items (30.9 per cent of all tariff lines), most of seed materials, breeding animals, components of feeding stuff for animals, as well as for a larger number of products that are not produced in the country, or that are produced in small quantities -- olives, tropical fruit, oriental spices and teas, coffee not roasted, cocoa.
- Annex IIIb: products subject to gradual reduction of import duties to zero in year 6. The list contains 523 tariff items (21.3 per cent), less sensitive products that are produced in the country. They include live animals, fresh milk and sour dairy products, except for yogurt or cheese, fruit and flower seedlings, cut flowers, a large number of vegetable products and continental fruit, roasted coffee, seed maize, sunflower, sugar beet and fodder seed, hard wheat flour and meal, maize and barley meal, pellets and rolled cereals grains, maize and potato starch, soybeans flour and groats, malt, molasses, raw soybeans and sunflower oil, larger number of preserved products made of fruit and vegetables, juices made of tropical fruit, wheat meals, shredded sugar beet, oilseed cakes.
- Annex IIIc: products subject to gradual reduction to zero yet subject to seasonal duties also after the transition period (10 tariff items, mainly highly sensitive products: tomatoes and a set of fruits);
- Annex IIId: products subject to a reduction of customs duties according to a timetable, yet where duties remain even after year 6 (33 tariff items, such as wheat flour, meat of swine, sour and sweet cherries).

This means that a large number of tariff lines in nearly all major CN categories of agricultural products will keep customs protection at a certain though reduced level even upon termination of the implementation period. The following products remain excluded from concessions: edible sunflower oil, raw and refined sugar, isoglucose, fructose syrups, wines (except from a customs free import of 25,000 hectolitres) and tobacco.

Danilo *et al.* (2010) made a first preliminary assessment of the effects of the first year of the implementation of the trade agreement on the bilateral trade flows and the possible consequences for the near future. The authors' expectation is that the Serbia's preferential status will contribute to further growth of the country's agricultural export to the EU in the medium term. Recent growth of exports of fruit, vegetables and cereals, although largely the result of abundant harvests, could benefit from tariff reductions, while the authors also point at the strong and increasing export flows of sugar, vegetable oils and fats and beverages and tobacco to the EU. On the other hand, exports of live animals and animal products (meat, dairy, fish) may face difficulties, despite import tariff reductions, due to a lack of legal harmonization and delay in the introduction of necessary standards in the field of food safety, traceability and quality in the whole supply chain. The authors observe that the 2009 Serbian exports of live animals, meat and meat preparations were only half of the 2007 export values. Serbian export volume of baby beef to the EU was about 1,000 tonnes in 2009, only 11 per cent of the approved quota. Also, exports of dairy products were less than what could have been potentially, according to Tomic *et al.*, as they point at the fact that while several dairy plants fulfilled the technical requirements necessary to export to the EU at the beginning of 2009, this was not the case with primary producers. The authors, therefore, plead for increased efforts and budget support for investments that will enhance productivity and competitiveness of production, restructure and modernise the supply chain, accelerate development of market institutions and improve the functioning of agricultural markets.

Further modernization of the agri-food sector is also necessary to face the competition from the EU which will benefit from improved access to Serbia as part of the ITA. Tomic *et al.*, however, are not expecting a massive inflow of EU agricultural products because of the trade liberalization measures. They argue that only a limited number of products have free access (no duties anymore) from the start of the implementation of the agreement. These are not sensitive products, and neither are they the products listed in Annex IIIa; they are largely products the country does not produce enough of in order to match demand. For instance, imports from the EU are concentrated within a small number of tariff items that refer to seed and seeding material, feeding stuff for animals, coffee, fruit and vegetables, all products insufficiently produced in the country. Most pressures from competitive imports from the EU may be expected from pork, fruit juices, fresh apples, tomatoes and sweet peppers (Danilo *et al.*, 2010: 81-97).

Agricultural trade figures for 2010 show no significant impact of the ITA as yet (see figure on bilateral trade in section 8.1), or at least there has not been a major inflow of EU agricultural products into the Serbian market. Serbia's exports, however, have continued growing. Preferential market access given to Serbian sugar and baby beef has been fully exploited by the sugar producers; Eurostat reports 190,000 tonnes of sugar import from Serbia in 2010, which even exceeds the zero in-quota tariff of 180,000 tonnes. Export of baby beef, however, was around 1,000 tonnes in 2010 (a similar level as in previous years), which is far below the approved TRQ.

8.4 Serbia's agricultural trade profile

Serbia is a net-exporter of agricultural and food products, both to the world as to the EU-27. Serbia's pattern of self-sufficiency in agricultural products can be crudely summarized as over-supply in crop products and self-sufficient in animal products. Looking more into the details of the figures points out that Serbia's self-sufficiency is most pronounced in cereals, sugar and fruits (see Table 8.10). The review of the trade policy in section 8.2 suggests that the cereal exports as well as the internal market balance of the animal products are largely the result of significant border protection.

Table 8.10. Self-sufficiency ratios (%) 2001, 2005 and 2009.

Commodity	2001	2005	2009
Eggs	93	100	62
Honey	31	115	74
Wheat	126	109	133
Maize	118	131	144
Soybean	122	124	75
Sunflower	99	96	86
Sugar	116	167	175
Potato	n.a.	106	101
Apple	88	113	222
Sour cherry	158	203	656
Raspberry	n.a.	467	435

Note: Self-sufficiency = domestic production as percentage of domestic usage.
Source: GoRS (2011), chapter 11.

Border protection may allow domestic prices to be higher than international market prices, whereas export subsidies can be used to bridge the gap between domestic and international prices when there is a surplus. The comparison between Serbia and several neighbouring and EU countries of farm gate price levels of major agricultural commodities indicates, however, that Serbia's farm gate prices are relatively low, hence directing attention to the relatively limited price support offered to the Serbian agricultural sector (see section 4.3). This is confirmed by agricultural policy measurements, showing that market price support is only a minor instrument of Serbia's agricultural support system whereas direct payments as input subsidies are the main element of agricultural budgetary transfers.[1] The market and trade figures build a picture of a rather modest degree of policy-induced market distortion for most of the agricultural products, although for some products (both from the crop as well as from the livestock sectors) protection rates are rather high (20-30 per cent, plus specific rates) and may affect trade outcomes. The opening up of these products to competition from the EU would have a significant economic and social impact, whose consequences would have to be foreseen and managed by policy makers.

8.5 Conclusions

Serbia is a net importing and open economy with foreign trade as a significant part of GDP. The EU is by far the country's major trading partner, for all merchandise trade and for trade

[1] To measure the level of agricultural protection several concepts can be applied, like the WTO's Aggregate Measure of Support (AMS) or OECD's Producer Support Equivalent (PSE). So far, these concepts have not been used to estimate the agricultural support in Serbia. As an alternative the Agricultural Policy Measures (APM) database has been designed and applied to measure agricultural support in several EU candidate and pre-candidate countries (Rednak and Volk, 2010). This measure follows the principal criteria of the PSE measure.

in agricultural products. Serbia is a net exporter of agricultural products, especially of fruit and vegetables, cereals and sugar.

Serbia is close to the finalization of its negotiation for WTO membership. That membership would integrate the country further into the global economy. The compliance with WTO rules is well under way, although there are still some critical issues to be tackled with respect to import licensing, sanitary and phytosanitary issues, intellectual property rights and services. A consequence of becoming a member of the WTO is that Serbia's import tariffs on agricultural products will decline to offer more market access to trade partners, who will have to do the same. On the other hand, although current MFN import tariffs are still significant, they are of little or no effect in regard to trade with major trading partners due to several free trade agreements, most notably under the CEFTA and the Interim Trade Agreement as part of the SAA with the EU.

Many agricultural products traded between Serbia and the EU are free of tariffs and quantitative restrictions. Nevertheless, there are also a substantial number of products for which Serbia will apply TRQs and/or tariffs in the years to come while there are also categories of products for which import duties are subject to gradual reduction over a period of 6 years. This implies that for a substantial group of products customs protection will remain at least for some years. Serbian agricultural exports to the EU are free of any duty or levy, except for baby beef, sugar and wine, which are subject to TRQs. Except for these categories mentioned, Serbian products have free access to the EU, but whether they will use the opportunities offered depends on their competitiveness in terms of price and quality. In addition, exports may face difficulties due to a lack of legal harmonization and delay in the introducing of necessary standards in the field of food safety, traceability and quality in the whole supply chain. To date Serbian standards and regulations are not yet fully harmonized with those of the EU.

Literature and websites

Bogdanov, N. and D. Bozic, 'Review of agriculture and agricultural policy in Serbia.' In: *Agriculture in the Western Balkan countries.* T. Volk (ed.), IAMO, Halle, Studies on the agricultural and food sector in Central and Eastern Europe, Vol. 57, 2010.

Danilo, T., V. Popovic and B. Vlahovic, 'Implications of the interim trade agreement implementation on agrarian economy in Serbia.' In: D. Tomic and M.M. Sevarlic (eds.), *Agriculture in late transition. Experience of Serbia.* DAES, Chamber of Economy of Vojvodina. Belgrade, 2010.

GoRS, *Answers to the European Commission's Questionnaire, Annex 30 to Chapter 29, Customs union in the EC.* 2011.

GoRS, *Answers to the European Commission's Questionnaire.* Chapter 11 Agriculture and Rural development. 2011.

ITA, *Interim agreement on trade and trade-related matters between the European Community of the one part, and the Republic of Serbia, of the other part, CE/SE/INT.* Available at: http://ec.europa.eu/enlargement/pdf/serbia/key_document/interim_agreement_trade_en.pdf.

Rednak, M. and T. Volk, 'Agricultural policy measures template. A tool for classifying and analysing agricultural policy measures' In: T. Volk (ed.), *Agriculture in the Western Balkan Countries.* IAMO, Halle, Studies on the agricultural and food sector in Central and Eastern Europe, vol. 57, 2010.

World Bank, 'Serbia Trade Brief.' In: *World Trade Indicators 2009/10: Country Trade Briefs.* World Bank. Washington, DC, 2010. Available at: www.worldbank.org/wti.

WTO, *WTO News, Serbia enters final accession lap.* 20 October, 2010. Available at: www.wto.org.

Chapter 9

9 Environment and agriculture

9.1 Introduction

This chapter studies the relationship between agriculture and the environment in Serbia. The justification is twofold. The first is the close relationship between agriculture and the natural environment that sustains this activity. The main function of agriculture is the production of food and other primary goods, but farming activities also influence positively or negatively the natural environment. Agriculture uses natural resources such as land and water, and the degradation or even exhaustion of these resources and biodiversity loss are classic examples of negative externalities of agricultural activities with negative consequences for the continuation of production. On the other hand certain extensive farming practices support impressive levels of agro-biodiversity. In Serbia agriculture's impact on the environment has been both positive and negative, with extensive agriculture contributing to valued cultural landscapes, while agricultural intensification and abandonment have adverse effects.

The second reason is that environmental targets and criteria are an important part of agricultural policy in the EU. The environment has started to receive growing attention in Serbia, particularly since 2004 with the introduction of the new legal framework for environmental protection (i.e. Law on Environmental Protection) and establishment of the Serbian Environmental Protection Agency (SEPA).

Section 9.2 describes the environmental impacts of agriculture. Section 9.3 examines the extent to which agri-environmental concerns are dealt with by the environmental, agricultural and rural policies. Section 9.4 discusses how institutions function to enforce the agriculture and environmental legislation. Section 9.5 describes the current state and prospects of organic farming. Section 9.6 deals specifically with biodiversity in Serbia and related institutional developments. Finally, the key conclusions are reported in section 9.7.

9.2 Environmental impacts of agriculture in Serbia

9.2.1 Impacts on water resources

Water consumption

Most of Serbia's country territory lies within the Danube Basin. Serbia's total fresh water resources[1] in 2009 were around 176,000 million cubic metres or 24,000 cubic metres per

[1] Fresh water resources refer to the total volume of water available in a territory, resulting from internal flow (water from precipitation minus evapotranspiration in a territory) as well as external inflow (water inflow from neighbouring territories) (Eurostat).

inhabitant (Eurostat, 2011). The LTAA (Long term annual average, >20 years) is estimated at 175,376 million cubic metres. According to Eurostat, approximately 96 per cent of all Serbia's fresh water originates from outside its territory. International cooperation on water issues is thus vital for Serbia.

The water resources of Europe as a whole are sufficient to meet water requirements. Comparisons of total fresh water abstraction with the available resources by country suggest that most European countries have enough resources to meet water demand; however, this percentage varies widely among the countries (EEA and WHO, 2002). The total water available for use within the Serbian territory or net abstracted water (data only available for total surface and groundwater) remains low, at around 2 per cent of fresh water resources. For comparison purposes, neighbouring countries like Hungary and Romania abstract 14 per cent and 10 per cent respectively of their potential and European southern countries like Spain and Greece abstract 29 per cent and 13 per cent, respectively (Eurostat, 2011).

Nevertheless, water use in Serbia is increasing. Data from Eurostat show a growth of 34 per cent in water abstraction over the period 2002-2009, reaching a total of 4,121 million cubic meters or 562 thousand litres per capita in 2009. Approximately 87 per cent of it came from fresh surface waters and the rest from ground water. Yet, in Serbia only approximately 2 per cent of total fresh water abstracted in 2009 was used for agriculture, against 81 per cent for the industrial sector. For comparison purposes, in countries like Spain (60 per cent), Denmark (20 per cent) or Romania (17 per cent) agriculture is a much bigger user of water than in Serbia.

As regards agricultural use of water, in 2008 approximately 89,000 hectares (FAOSTAT, 2011) out of 5 million hectares of agricultural land in Serbia were equipped for irrigation, which constitutes 2 per cent of agricultural land.[1] In 2010 the total water abstracted for irrigation was 65 million cubic meters, and the total irrigated area was 25,128 ha, which is 29 per cent of the area covered by irrigation systems, while in relation to total agricultural area the share of the irrigated area was 0.5 per cent. The most frequent type of irrigation was by spray irrigation (89 per cent of area irrigated by spray irrigation, 5 per cent irrigated by surface irrigation, while drop-by-drop irrigation represented 6 per cent of the irrigated area). The total irrigation waters used have declined from 90 million cubic metres in 2006 to 65 million cubic metres in 2010, and the facilities and equipment for irrigation have also declined since then (Statistical Office of the Republic of Serbia, 2011).

The following Table 9.1. depicts the regional distribution of the irrigated area in Serbia. The irrigated area is concentrated in Belgrade and Vojvodina.

[1] In contrast, some 30 per cent of agricultural land is equipped for irrigation in countries like Greece and Italy, and in Spain this percentage is almost 15.

Table 9.1. Irrigated area under crops/plantations, 2010 a).

	Total Arable Area ha	Irrigated Area ha	%
Total	5,091,507	25,128	0.5
Belgrade	217,374	1,583	0.7
Vojvodina	1,589,203	14,972	0.9
Sumadija and Western Serbia	1,500,578	329	0.0
Southern and Eastern Serbia	1,784,352	8,244	0.5

Note: a) Preliminary data.
Source: Statistical Office of the Republic of Serbia (2011).

Water quality

Water protection is still an underdeveloped segment of the water management system in Serbia (MAFWM, 2011). The absence of control of water quality used for irrigation may be a cause of water pollution, although only 2 per cent of agricultural land is equipped for irrigation. Agriculture mainly contributes to the pollution of waters from diffuse sources of pollution, which in the watershed of the Danube gives 70 per cent of total nitrogen, 50 per cent of total phosphorus and 90 per cent of faecal and coliform bacteria. The main agricultural sources of water pollution are the uncontrolled effluents of livestock farms (e.g. animal manures), crops, slaughterhouse wastes and chemical fertilizers. However, it is difficult to estimate to what extent agriculture contributes to water pollution in Serbia, because there is no complete nor effective monitoring of ground water quality (Bogdanov, 2007).

Drainage

The used agricultural area covered by a drainage system in 2010 was 1,971 thousand ha, while the used agricultural drained area was 1,673 thousand ha, representing an area of 33 per cent of total agricultural land. According to the Serbia country review (World Bank, 2007), problems with poor drainage have led to waterlogging, salinization and erosion. Drainage channels, associated structures and pumping stations have deteriorated over time. Rehabilitation of the structures is required. Estimations presented in the same study show a 20 per cent to 30 per cent crop yield increases as a result of the improvement of drainage systems. There are no recent data available to evaluate the current drainage system situation.

Flood protection

The Ministry of Agriculture, Forestry and Water Management (MAFWM, 2011) provides some information on the built flood protection systems, which account for 3,460 km of dykes, 820 km of regulated waterways, 39 accumulation and retention ponds, and 930 km of multipurose canals as part of the Danube-Tisa-Danube hydro system. The flood protection system that was built is still inadequate, however, therefore activities for constructing additional dykes, and reconstructing existing ones, are under way. The area of eroded land in 2010 amounted to 2,623 km^2, which represents 18 per cent less than in the previous year (SORS, 2011).

Wastewater management

Wastewater is water which is, after use, treated as wastewater and transported to a treatment plant or let out into a recipient (underground or surface water) (SORS, 2011). In the Republic of Serbia wastewater from households has been increasing slightly since 2008 to 2010, while wastewaters from industry and other uses (including agriculture) have declined in the same period. In 2010, 60 million cubic metres, or 17 per cent, of the 350 million cubic metres of waste water have been treated. This percentage is below the EU average.

9.2.2 Use of pesticides and fertilizers

A considerable portion of Serbia's area is devoted to agricultural production. According to FAOSTAT[1] data, in 2008 57 per cent of total land area was dedicated to agriculture (FAOSTAT, 2011). The development and intensification of agriculture has consequences for quality of soils and for both quantity and quality of water. Concerns about the impact of agriculture on the quality of water are mostly related to the leaching of agricultural chemicals applied to crops and soil. Fertilizers and pesticides are the most important sources of water pollution. The consumption pattern of these agricultural inputs varies among the different elements.

While in Serbia the consumption of insecticides, fungicides, bactericides and other pesticides shows a clear declining trend since the beginning of the 1990s, the consumption of herbicides shows a more stable trend since 1995 (Figure 9.1). Data on the consumption of pesticides in Serbia after 2004 could not be found.

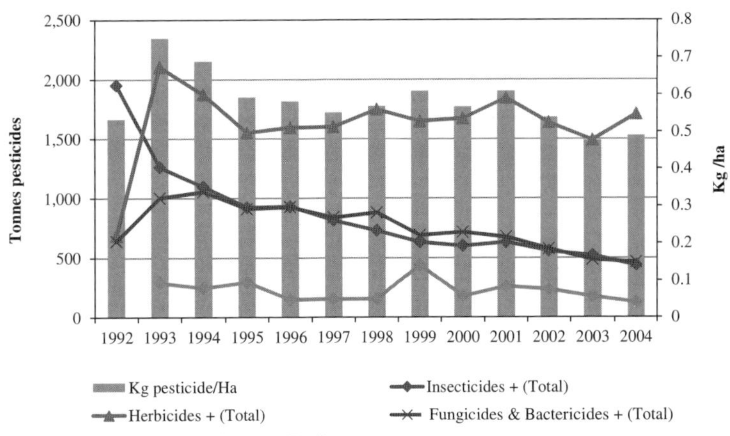

Figure 9.1. Consumption of pesticides in Serbia and Montenegro, 1992-2004. Source: FAOSTAT (2011).

[1] Statistics Division of the Food and Agriculture Organization of the United Nations.

Figure 9.2 shows that in 2004 agriculture in Serbia used 0.5 kg/ha of pesticides. This is relatively little and comparable to levels recorded in Romania and Estonia. The use of pesticides (in kg/ha) is more than ten times higher in countries like the Netherlands, Italy and Belgium (FAOSTAT, 2011).

Figure 9.2 shows the consumption of the three types of fertilizers between 2002 and 2008 in Serbia. The level of consumption of phosphate and potash fertilizers measured in tonnes of nutrients remains stable during the period and is lower than the consumption of nitrogen fertilizers. The latter shows a significant decline in 2008 after rebounds of consumption in 2004 and 2007. Serbian consumption of nitrogen fertilizers was about 22 kg/ha in 2008, which is below levels in neighbouring countries like Croatia (30 kg/ha) and Hungary (33 kg/ha) and significantly less than the use in Poland (48 kg/ha) and in the Netherlands (62 kg/ha) (FAOSTAT, 2011).

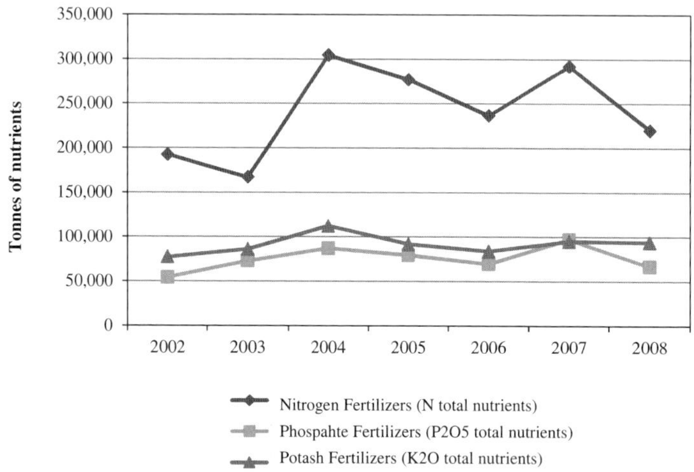

Figure 9.2. Fertilizer consumption in Serbia a) (2002-2008), tonnes.
Note: a) 2002-2006 data for Serbia & Montenegro.
Source: FAOSTAT (2011).

Eutrophication is the process by which a body of water acquires a high concentration of nutrients, especially phosphates and nitrates. A study from the Serbian Environmental Protection Agency (SEPA, 2007) on the concentration of nutrients in rivers,[1] based on data on water quality from the Republic Hydro Meteorological Service in Serbia (Republic Hydro Meteorological Service of Serbia), shows that during the 1980s, while agricultural production was increasing, the concentration of nutrients also increased. During the 1990s the economic crisis contributed to the reduction of pollution from agriculture. The study

[1] Concentration of nutrients in rivers is taken as indicator for the concentration of nutrients in fresh water in this study because the results for lakes and groundwater have to be taken with precaution because of the inconsistent time series (SEPA, 2007).

concludes that the slow growth of agricultural production during recent years has not added a significant pressure on the rivers in comparison to the values observed in the 1980s.

However, the water monitoring as foreseen by Directive 91/676/EEC (Nitrates Directive) has not been established yet in Serbia, especially for the nitrate vulnerable zones, and the nitrate vulnerable zones in the Republic of Serbia have not yet been designated (GoRS, 2011). Furthermore the Code of Good Agricultural Practices on farms has been created in Serbia in 2008 in the form of a brochure, but the 'Code' is not mandatory for the time being and the regulation for the application of it does not exist yet. The implementation of the Nitrate Directive in Serbia has not yet been initiated.

According to the Report on the State of the Environment 2010 on the Vojvodina Province (area marked by highest pollution pressure), high phosphorus content was found in 25 per cent of analysed soils and 56 per cent of the territory is in danger of further acidification (Cooper *et al.*, 2010).

9.2.3 Impacts on the soil

The soils of Serbia are very heterogeneous due to the varied geological base, climate, vegetation and fauna. Soils are divided in 8 fertility classes reflecting the suitability for agriculture production. Classes I to IV represent the higher quality soils. In Serbia 45.8 per cent of total land area belongs to classes IV to VIII, which are less suitable for agriculture production.

Table 9.2. Land quality classes in Serbia.

Soil class	km^2	%
I	11,650	13.2
II	9,357	10.6
III	10,522	11.9
IV	8,682	9.8
Arable	40,211	45.5
V	11,073	12.5
VI	20,144	22.8
VII	8,069	9.1
VIII	1,178	1.3
Non-arable	40,464	45.8
Productive	80,675	91.3
Infertile	7,686	8.7
Total	88,361	100.0

Source: SEPA (2007).

The major threats to soil identified in the literature for Serbia and presented in order by intensity of threat are: soil sealing, decline in soil organic matter, soil acidification, soil contamination and soil erosion (Antić-Mladenović and Ličina, 2010). Approximately 88 per cent (or 7.7 million ha) of the total Serbian land is affected by water erosion. Wind erosion affects approximately 25 per cent (or 2 million ha) of the Serbian territory, of which 18 per cent corresponds to the province of Vojvodina. The north is mainly affected by wind erosion while the south is more impacted by water erosion.

Table 9.3 shows the distribution of categories of water erosion in Serbia. About half of the Serbian territory suffers light or very light erosion, while only 3 per cent of the total land is affected by severe erosion.

Table 9.3. Estimated distribution of water erosion in Serbia (based on erosion categories).

Degree of erosion	Area (1,000 ha)	Share of total area (%)
Very light	1,165.2	13
Light	3,933.4	45
Medium	1,633.2	18
Strong	774.1	9
Extreme	241.5	3
Total	7,747.4	88

Source: Antić-Mladenović *et al.* (2010).

Agriculture inevitably lowers soil fertility by removing soil nutrients incorporated in the harvested crops. Excessive cropping or grazing can reduce soil-nutrient levels and degrade soil structure.

Artificial fertilization in agriculture contributes at the same time to higher yields and to eutrophication of aquatic ecosystems. Both effects need to be considered by finding the optimal use of artificial fertilizers.

Soil acidification decreases the rate of humidification. Soil salinization is a process of accumulation of readily soluble salts in parts of soil profiles. Inadequate irrigation may speed up salinization, a process that is more evident in Vojvodina than in any other part of the country.

Exposure of soils to wind and rain during cultivation contributes to erosion of the fertile surface. The law on Agricultural Soil of 2006 establishes several measures to prevent erosion, such as an obligation to apply crop rotation, change the tillage system, plant shelterbelts and a ban on felling forest (Antić-Mladenović and Ličina, 2010). However, it is unknown how effective this law has been to date.

9.2.4 Impacts on climate change

According to data from the World Bank, agricultural methane emissions in Serbia in 2005 represented 44 per cent of the country's total methane emissions, or 3,400 thousand metric tonnes of CO_2 equivalent; and agricultural nitrous oxide emissions had a share of 64 per cent of total nitrous oxide emissions, or 2,914 thousand metric tonnes of CO_2 equivalent. For comparison purposes, in the EU agricultural methane emissions and agricultural nitrous oxide emissions represent 41 per cent and 56 per cent, respectively, of these emissions from all sources.

Data from UNFCCC (United Nations Framework Convention on Climate Change) from 1998 attributed to the agricultural sector a share of 14 per cent of total Greenhouse Gas (GHG) emissions. According to this source (UNFCCC, 2011), agricultural soils accounted for 54 per cent of the agricultural emissions, livestock enteric fermentation accounted for 30 per cent and manure management accounted for another 14 per cent. It has not been possible to find data on specific emissions of CO_2 generated from the combustion of fossil fuels in agriculture. However, data from FAOSTAT show a reduction in the number of tractors and agricultural machinery in the last decade.

9.2.5 Concluding remarks on agriculture's environmental impacts

In Serbia the impacts of agriculture on the environment are both negative and positive. Agricultural intensification brings on adverse effects, while extensive agriculture contributes to highly valued cultural landscapes.

Water

In 2009 the total net abstracted water in the Republic of Serbia was 4,121 million cubic metres, representing 2 per cent of total fresh water resources. For comparison purposes, Hungary and Romania abstract 14 per cent and 10 per cent, respectively, of their potential. In Serbia approximately 2 per cent of the fresh water abstracted in 2009 was used for agriculture, and the share is declining since 2002. Out of the 5 million hectares of agricultural land, 2 per cent were equipped for irrigation; however, only 0.5 per cent of total agricultural land was irrigated. Total used irrigation water has declined, and facilities and equipment for irrigation have also declined since 2006.

Water protection is still an underdeveloped sector of the water management system in Serbia (MAFWM, 2011). The main agricultural sources of water pollution are the uncontrolled effluents of livestock farms, crop and slaughterhouse wastes and chemical fertilizers and crop protection compounds. It is difficult to estimate to what extent agriculture has an influence on the pollution of waters in Serbia because of the lack of an efficient monitoring system; however, Serbia is one of the largest contributors of nitrogen and phosphorus to the Danube River. Studies in this field indicate that the construction of proper manure storage facilities and manure spreading equipment are in a rudimentary stage; however, these measures as well as the rationalization of the use of mineral fertilizers are part of the Danube River Enterprise Pollution Reduction project (DREPR). Rehabilitation of drainage structures is required and the flood protection system that was built is inadequate. Approximately 17 per cent of waste water is treated; this percentage stays below the EU average. Summarizing, a large investment effort seems to be necessary in Serbia in order to achieve an adequate water management system.

Pesticides and fertilizers

Data on the consumption of pesticides in Serbian agriculture could not be found after 2004. Until then a declining trend in their consumption could be observed. In 2004 Serbian consumption of pesticides per ha of agricultural land (0.5 kg/ha) was relatively small compared to other European countries. Between 2002 and 2008 the level of consumption of phosphate and potash fertilizers measured in tonnes of nutrients remained stable, while consumption of nitrogen fertilizers declined after a rebound in 2004. A study of the Serbian Environmental Protection Agency (SEPA) of 2007 concludes that in the recent years the slow growth of agriculture production has not added significant pressures on the rivers of Serbia.

However, the implementation of the Nitrates Directive (Directive 91/676/ECC) of the EU has not been initiated yet and the water monitoring system as foreseen in this Directive has not been established yet either. Furthermore, the Code of Agricultural Good Practices has been created in Serbia in 2008 as a brochure, but it is not mandatory and the regulation for its application is not available.

Impacts on the soil

In Serbia 45.5 per cent of total agricultural land is very suitable for agricultural production (class I-IV). Agriculture inevitably lowers soil fertility by removing soil nutrients in the harvested crops. Excessive cropping or grazing can reduce soil-nutrient levels and degrade soil structure. Exposure of soils to wind and rain during cultivation contributes to erosion of the fertile surface. The law on Agriculture Soil of 2006 established several measures to prevent this type of erosion, but its effectiveness is unknown.

Impacts on climate change

According to World Bank data, the agricultural methane emissions in Serbia in 2005 represented 44 per cent and the agricultural nitrous oxide emissions 64 per cent, respectively, of the total of these emissions from all sources. For comparison purposes, the share of EU agricultural emissions for these components is 41 per cent and 56 per cent, respectively.

9.3 Overview of policies relating to the environmental impacts of agriculture

9.3.1 Environmental policy

EU environmental policy aims at promoting sustainable development and to protect the environment for present and future generations. The compliance with the EU's *acquis*, which comprises legal acts covering air quality, waste management, water quality, nature protection, industrial pollution control and risk management, chemicals, climate change, noise and civil protection, requires significant investment.

Serbia's Constitution (Official Gazette of RS issue 98/2006) guarantees environmental protection and sustainable development. The framework law in the area of environmental protection in Serbia is the Law on Environmental Protection, which was adopted in 2004 and amended in 2009. The law contains basic principles that regulate the protection of the environment. The law also regulates management of natural values and conditions for environmental protection, monitoring the state of the environment, public information and public participation, economic instruments, liability for pollution, supervision and penalty provisions (GoRS, 2011). Under this framework law the Republic of Serbia created special laws for environment protection, several of which have direct relevance for agriculture, such as the Law on Air Protection,[1] Law of Nature Protection,[2] the Law on Environmental Impact Assessment[3] and the Law on Integrated Prevention and Control of Environment Pollution.[4] The principles that underpin the environmental legislation are the integration principle, prevention, natural value preservation and sustainable development, polluter's liability, principles of 'polluters pay' and 'users pay', subsidiary liability of the state authorities when the polluter is unknown or damage is caused by pollution originating from outside the Republic of Serbia, principle of incentives, public information and participation and protection of rights to healthy environment and access to justice.

In May 2008 the National Sustainable Development Strategy (NSDS) for the period 2009-2017 was adopted. Furthermore an institutional framework for its implementation has been set up, as well as sources of financing and a mechanism of monitoring with clearly

[1] Official Gazette of RS, No. 36/09.
[2] Official Gazette of RS, Nos. 36/09 and 88/10.
[3] Official Gazette of RS, Nos. 98/2002 and 36/09.
[4] Official Gazette of RS, No. 104/08.

defined indicators, which comply with the revised list of UN sustainable development indicators. The NSDS also complies with the EU sustainable development strategy, Millennium Development Goals (UN) and with a National Strategy for Accession of Serbia-Montenegro to the European Union (adopted in June 2005). The Action Plan for the implementation of the NSDS for the period 2009-2017 was adopted in May 2009. The Action Plan contains specific measures and activities for the implementation of the NSDS, and determines time limits, total costs and sources of financing. In April 2010 the Action Plan was amended by including indicators for monitoring the implementation of measures.

In January 2010 the National Environment Protection Programme (Official Gazette of RS No. 12/10) was adopted by the Serbian Government. This programme contains short-term (2010-2014) and long-term objectives (2015-2019). The Programme aims at increasing environment quality and improving living quality of the population in the Republic of Serbia. As a strategic document, the Programme stands as an instrument establishing the priorities in the field of environmental protection in the country. According to official sources, (GoRS, 2011) the Programme contains a description and estimation of the state of the environment, measures of protection and implementation methods, timelines and budgets. A report on the implementation of the Programme needs to be prepared and submitted by the Ministry every 2 years to the Government. The Action Plan is being prepared and it is expected to be completed in 2011.

The alignment of the Serbian environmental legislation and policies with the EU legislation is part of the National Programme for Integration of the Republic of Serbia into the European Union (NPI), adopted by the Government of the Republic of Serbia in October 2008. The NPI contains an overview of the duties that must be realized by 2012, which apply to legislative actions (transposition and implementation of EU regulations into national legislation), strengthening of institutional and administrative capacities and stating financial needs. There are several technical projects which assist the approximation of legislation and policies, like the cooperation within the International Commission for the Protection of the Danube River. EU and World Bank experts engage in these projects providing advice aimed at improving laws. The Directorate General for the Environment (DG Environment) of the European Commission initiated in 2006 the Progress Monitoring Project with the objective to assist DG Environment and the beneficiary country in monitoring, assessing and documenting the progress in environmental law harmonisation and implementation process. With this project DG Environment aims at producing a progress report based on tables of concordance and implementation questionnaires (Simić, 2007).

In February 2011 the Biodiversity Strategy and Action Plan of the Republic of Serbia has been adopted. The Biodiversity Strategy has been developed by the MEST and the United Nations Development program (UNDP). The legal basis for the adoption of this strategy in the Republic of Serbia is the Law on Government, in accordance with the Law on Ratification of the Convention on Biological Diversity.

At the international level 73 international agreements in the field of environmental protection have been ratified, and the ratification of other international agreements is pending (GoRS, 2011).

9.3.2 Agricultural policy

The first Agricultural and Rural Development Strategy of the Republic of Serbia, adopted in 2005, defined several general strategic goals with direct impact on environmental and sustainability concerns, i.e.:

- Build up a sustainable and efficient agricultural sector;
- Ensure support to sustainable rural development;
- Protect the environment from the negative effects of agricultural production.
- The Law on Agriculture and Rural Development (Official Gazette of RS, No. 41/09) is part of the current basic framework of agricultural policy and regulates, inter alia, the objectives and how to achieve them, types of incentives and conditions for beneficiaries, Agrarian Council, supervision and other issues of relevance to agriculture and rural environment. The agricultural incentives are divided into three groups: direct payments and support (with the objective of reducing the cost of agricultural products and facilitating an increase of supply), market (price) incentives and structural incentives (aiming at supporting activities that contribute to sustainable modernisation of agriculture) (see Chapter 7).

In May 2010 the National Parliament of the Republic of Serbia passed a new set of laws and amendments to the laws related to agriculture proposed by MAFWM. The new set of laws is expected to contribute to the improvement of overall environment for agriculture production, to the achievement of compliance with international rules as well as to further harmonisation with EU laws (Maslac, 2010). Several laws of the new set have a direct relation with the environment:

- Law on Forests which defines a system of forest protection, management of forests and obligatory registration of all forests. A forest fund will be set up that will be partly financed from fees from the use of forests and forest land.
- Law on Waters which include charges to legal entities for water use, release, pollution, and drainage, the use of water supply facilities and systems and sewer water. It also includes fees to provide incentives for the building of irrigation systems. The law stipulates that companies must pay 0.025 per cent of their revenues into an irrigation fund (Maslac, 2010).
- Amendments to the Agriculture Extension and Advisory Service, which improves the regulated advisory system aimed at informing farmers about modern standards of products quality, farm management, and environmental protection.
- In the support to agricultural production in Serbia, cross-compliance measures with environmental, animal welfare, public, animal or plant health standards do not exist. However, the basis for the entitlement to use the incentive funds of MAFWM is the compliance with legal norms relating to agro-environmental measures. The latter will form the basis for the future implementation of a cross-compliance system. From 2011, within the scope of the preparation for IPARD (Instrument for Pre-Accession Assistance for Rural Development), the obligation of respecting minimum national standards shall be introduced (GoRS, 2011, Chapter 11: 49).

Input subsidies are generally controversial and do not take into consideration the long-term sustainability of soil resources. In Serbia the agricultural budget does contain input subsidies, yet these are classified as reimbursement of costs of inputs like fertilizers and fuels.

9.3.3 Rural development policy

In the past, the Rural Development policy was not defined in a specific programme in Serbia but as a part of the agricultural policy of the country. Among the rural development measures, several may have a direct impact on environment such as the support for promotion of organic production, for sustainable use of genetic resources of domestic animals and for sustainable use of plant genetic resources with regard to food and agriculture, all which are in the Pillar 2 of the CAP.

9.4 Environmental institutions and their functioning

The significance of environmental issues was recognized during the drafting of the National Environment Protection Programme, which has been developed on the basis of the provisions of the EU's 6th Environmental Action Plan.

The first general objective of environment protection policy, pursuant to the National Environment Protection Plan, is the integration of environment protection policy with economic policy and the policies of other sectors.

9.4.1 Institutional framework

Based on the Law on Ministries (Official Gazette of RS No. 2008/65), duties in the field of environment, on the national level, fall under the jurisdiction of various institutions. The Ministry of Environment and Spatial Planning (MESP) is the first responsible institution for environmental and natural resource protection, with the Environmental Protection Agency as its administrative body, performing duties related to monitoring, registration, and policy implementation. In the field of environmental protection the Ministry of Agriculture, Forestry and Water Management (MAFWM) has a major responsibility for issues related to genetically modified organisms, water management and forestry.

Certain jurisdictions in the field of environmental protection were transferred to the Autonomous Province of Vojvodina in 2002 by adopting the Law on Determining the Jurisdiction of the Autonomous Province (AP) of Vojvodina (Official Gazette of RS, No. 6/02). Through the Provincial Secretariat for Environmental Protection and Sustainable Development, AP Vojvodina is responsible for affairs related to policy development and implementation of measures aiming at environmental protection and sustainable development within the territory of the province. The Province is also responsible for strategic assessment of plans and programmes and for issuing integrated permits for facilities and activities within the territory of the province.

According to the Budget Law of the Republic of Serbia for 2010, the distribution of the budget dedicated to environmental protection purposes is indicated in Table 9.4.

Table 9.4. Environmental Budget of the Republic of Serbia.

Institution	Euro	%
Ministry of Environment and Spatial Planning	16,368,072	84
Environmental Protection Agency	454,958	2
Chemicals Agency	677,083	3
Agency for protection from ionizing radiation	260,416	1
Fund for Environmental protection	245,437	1
Nature Protection Institute	1,593,750	8
Total	19,599,716	100

Source: GoRS (2011).

For comparison purposes, the environmental budget represents 0.2 per cent of the total budget of the Republic of Serbia, while the agriculture budget accounts for 2.6 per cent.

9.4.2 Policy implementation, monitoring and enforcement

Serbia has demonstrated its interest in and commitment to the protection of the environment through the ratification of international agreements and the adoption of a national legal framework since 2004, and in 2010 by the adoption of the National Environment Protection Programme. The Biodiversity Strategy has been adopted in February 2011. However some EU directives in the field of the environment (e.g. Nitrates Directive) have not been implemented yet. Serbia is having difficulties in implementing policies and strategies for adequate environmental protection. Besides a lack of financial resources, the national legal framework is recent and partly still under development. Central collection of data and appropriate monitoring of environmental issues are missing or fragmented. Enforcement and public awareness of the importance of the environment are open issues.

The major sources of the financing of environmental protection are the state budget and incomes from fees. The resources can also be provided by donations, credits, international assistance, foreign investments dedicated to environment protection, EU, UN and international organization instruments, programmes and funds.

9.4.3 Financing environmental projects

Investments and regular expenses of environment protection are all expenses that prevent, remove or decrease harmful impact to the environment (without investment expenditures and depreciation). According to data of the Republic Statistical Office, investments and regular expenditures for the period 2006-2008 are presented in the following table.

Table 9.5. Investments and regular expenditure of environment protection (in million dinars and, between brackets, in thousand euro).

		Waste disposal	Surface water protection	Air protection	Underground water and soil protection	Protection of nature	Protection against noise
2006	Investments	770.6 (9,754)	122.8 (1,555)	734.5 (9,298)	14.2 (180)	373.6 (473)	61.3 (776)
	Regular expenditures	153.7 (1,945)	49.9 (61)	21.8 (276)	18.7 (236)	88.6 (1,120)	0.7 (9)
2007	Investments	314.5 (3,956)	707.3 (8,896)	1 568.7 (19,732)	164.6 (2,070)	68.2 (857)	165.5 (208)
	Regular expenditures	334.3 (4,204)	65.4 (822)	86.1 (1,082)	82.7 (1,041)	149.0 (1,874)	3.1 (38)
2008	Investments	574.8 (6,459)	685.1 (7,698)	349.3 (3,925)	113.5 (1,275)	29.3 (329)	26.0 (292)
	Regular expenditures	578.5 (6,500)	132.9 (1,494)	150.5 (1,691)	76.8 (863)	375.0 (863)	5.0 (57)

Source: SEPA (2007).

9.5 Organic agriculture and relevant institutions

According to the EU definition, organic farming involves production of crops and livestock that favours cultural, biological and mechanical management methods rather than the use of off-farm inputs such as synthetic fertilizers and pesticides. Emphasis is placed on achieving environmental protection; and with regard to livestock production, on animal welfare considerations (European Commission, 2001). This section seeks to provide insights on the market for Serbian organic agriculture, its regulatory framework and institutional set-up and the prospects for the future of organic agriculture in Serbia. However, difficulties were encountered in the development of this chapter regarding the availability of data. Under the Law of Organic Agriculture and Organic Products (Official Gazette of RS No. 62/06) a single register was introduced for organic producers and processors, and traders in organic products. Nevertheless the number of farms and land in Serbia on which organic production methods are applied are larger than those registered, because the Ministry (MAFWM) does not have data on producers whose production is certified by control bodies that are not authorised by the Ministry (e.g. production dedicated to exports).

9.5.1 Market for organic agriculture

Serbia started producing certified organic products in late 1970s. Only for the last decade, production and processing of organic products became more popular and economically more important. Arable land under organic production in Serbia increased fivefold in 2009 compared with 2008 (MAFWM, 2011).

Organic development in Serbia is export driven. While vegetables and cereals are both produced for export and for domestic consumption, the products from the fruit sector (wild and cultivated fruits and berries) are mainly exported. Vegetable and cereal production is located mainly in the region of Vojvodina, while production of fruits and mushrooms is concentrated in Central and South Serbia (USDA, 2009).

In the absence of a central database system, it is hard to provide figures regarding organic production in Serbia and data vary between sources. USDA (2009) indicates that according to unofficial data Serbia's organic production takes place on approximately 5,000 ha, and an additional 9,000 ha of land would be in the period of transition towards organic production. That makes a total of organic production on some 15,000 ha or only 0.3 per cent of the total arable land in Serbia. Only 600 ha are certified by the organizations that are accredited by the Ministry of Agriculture, and Serbia has around 218 certified organic farmers. In the spring of 2010 the GIZ (Deutsche Gesellschaft fuer Internationale Zusammenarbeit GmbH) together with the AFC/FIBL[1] consortium conducted an organic sector survey aiming at collecting data inter alia on the area, crops, number of operators and regional distribution in Serbia. The survey suggests that in Serbia the arable land under organic farming accounts for about 8,500-9,000 ha and at least 3,000 small-scale farmers are involved in organic production (GIZ, 2011). The SEPA (SEPA, 2007) provides a much higher figure of certified organic area, approximately 200,000 ha. However this figure includes the area dedicated to organic wild fruits and berries, which accounts for 180,000 ha.

USDA estimated the total organic production in 2009 at about 30,000 tonnes of organic products, with a total market value of organic products in Serbia of about $55 million (38 million euro), while the GIZ study estimates the farm gate value between 20 and 25 million euro.

At present, more than 90 per cent of all organic production is exported. Serbia has a potential for growth in this sector, but Government support for organic farmers is limited due to the reduced agriculture budget for organic production (94,800 euro in 2010) (GoRS, 2011). The local market for organic food in Serbia is underdeveloped. According to the GIZ study not many certified organic products are imported. The imports are dominated by organic baby food, as Serbian organic baby food is not yet certified. The GIZ study suggests that the further development of the local market for organic products in Serbia requires a strong campaign with a clear communication strategy for consumers.

9.5.2 Institutional set-up and regulatory framework for organic farming

Organic production in the Republic of Serbia is regulated by the Ministry of Agriculture, Forestry and Water Management. As part of the Ministry, the Section for Organic Production supervises the performance of delegated tasks by authorized control bodies, keeps records of organic production, authorizes the use of conventional reproductive material and participates in granting exemptions from organic production rules. The Sector for Rural Development is in charge, inter alia, of the enhancement of the system of organic production, prepares the basis for new legislation, proposes measures for support and prepares information and analyses on the state of organic production. The General Inspectorate, which is part of the state administration within the Ministry, is responsible for the monitoring of the application of the law and other measures in the area of organic production. Control organizations are authorized by the Ministry and are independent entities that carry out the control of the application of the organic methods and issue certificates for products that are the outputs of organic production.

---

[1] FIBL: Research Institute for Organic Agriculture.

In Serbia organic production is regulated by the Law on Organic Production (Official Gazette of RS No. 30/10) in force since 1 January 2011. The predecessor of this law was the Law on Organic Agriculture and Organic Products (Official Gazette of RS No. 62/06). The new law aims at resolving deficiencies in the regulatory regime with regard to organic production and products in Serbia and provides the legal framework for tracking organic production, registration of production volumes, as well as measures to stimulate organic products (MAFWM, 2011).

A national organic label has been introduced though the rulebook on the national label for organic products. Every organic product that is properly certified in Serbia must have this label. The label is authorized and maintained by the Ministry of Agriculture. Exporting organic products can further complicate the certification process, because most nations have their own organic labels. In this case the producer must work with a certification body of the country in which the product will be marketed.

The Law on Organic Production (Official Gazette of RS No. 30/10) complies with the European Union regulations.[1] The National Plan for Organic Agriculture drafted in 2009 can be described as the Serbian version of the EU's Action Plan for Organic Food and Farming. Its overall and ambitious goal is to increase organic certified land to 50,000 ha by 2014.

9.5.3 Prospects for organic agriculture

According to USDA, Serbia has a potential for growth in this sector, but mainly in rural areas where conventional production is not possible or where sources of pollution are not nearby. Furthermore, during the past fifteen years most Serbian farmers could not afford synthetic inputs, which has resulted in low levels of prohibited residues and implies that in Serbia conversion periods[2] are shorter than in the most developed countries. On average the conversion in the EU can take up to 5 or 6 years, while in Serbia organic conversions within 1 or 2 years are not uncommon. This factor significantly enhances the potential for rapid growth in the organic sector in Serbia.

However, the potential for growth of this sector is limited by the reduced agricultural budget for organic production (94,800 euro in 2010) (GoRS, 2011). Subsidies for organic production for 2010 (Table 9.3) were about 94,800 euro (used for certification, education and marketing). In Serbia per hectare subsidies in 2010 for organic production oscillated between 237 euro/ha and 427 euro/ha (GoRS, 2011). For comparison purposes, in the EU-15 and the EU-10 per hectare payments for farms with organic production in 2007 were 438 euro and 324 euro, respectively (EC DG for Agriculture and Rural Development, 2010).

[1] Council Regulation (EC) No. 834/2007 on organic production and labelling of organic products and repealing Regulation (EEC) No. 2092/91, and Commission Regulation (EC) No. 889/2008 laying down detailed rules for the implementation of Council Regulation (EC) No. 834/2007.

[2] Before certification, farms must undergo a conversion period during which organic principles are implemented to reduce non-approved input residues in the fields (USDA, 2009).

Table 9.6. Subsidies foreseen in agrarian budget 2010 for organic production.

	Subsidies	RSD	Euro
Plant production, per hectare	Field crop production	25,000	237
	Vegetable production	35,000	332
	Fruit production and viticulture	45,000	427
Livestock production, per head	Large stock	15,000	142
	Small stock	5,000	47
	Poultry	500	5
	Beehives	2,000	19

Note: Exchange rate RSD/Euro 105.4982.
Source: GoRS (2011).

While the eco-climatic factors indicate a large potential for development of the organic sector in Serbian agriculture, the GIZ study suggests that the three main problems for the development of organic farming in Serbia are the capital constraint, the poor organization of agents along the food supply chain and low efficiency of production, processing and marketing. Suggestions to overcome these problems are improvements at the institutional level and the improvement of the agro-political concepts and financial instruments, education, promotion of modern technologies at the processor level and measures that allow for the exploitation of economies of scale at farm and production level, such as the propagation and support of associations and cooperatives.

9.6 Biodiversity

9.6.1 Biological diversity

Serbia is situated in the central part of the Balkan Peninsula in the South of Eastern Europe. The main climate type is mild continental. The rivers of Serbia end in the Black, Adriatic and Aegean seas. The Serbian country is highly mountainous in the central and south parts and mainly a plain in the north.

Biological diversity is the variety of life on earth and considers variety at all levels of biological organization from genes to species to ecosystems (Gaston and Spicer, 2004). Investigations of biological diversity have a long and rich tradition in Serbia (SEPA, 2007). The high biodiversity in Serbia is related to the variety of orographic, geological, climatic and historical factors. Nevertheless, biodiversity erosion is a process markedly impacting Serbia and the whole Balkan Peninsula. According to the study of the EPA of Serbia (SEPA, 2007), many species have disappeared in Serbia and some species are highly endangered mainly due to human induced pressures.

Ecosystem and habitat diversity

An ecosystem is a system formed by the interaction of a community of organisms within a certain space with their environment. In general there are two different approaches for the classification of ecosystems: the habitat[1]-oriented approach (similar environmental

[1] Habitat is the location or environment where an organism or a biological population normally lives or occurs.

conditions such as climate, hydrology, geology, soil) and a community-oriented approach (similar physiognomy or floral and faunal composition). At the EU level two directives specify the habitat-oriented classifications of ecosystems: the Directive on Wild Birds (EEC/79/409) and the Habitat Directive (EEC/92/43).

Ecosystem diversity can be presented by the number of plant communities. The vegetation in Serbia is extremely diverse with 700 to 800 associations or plant communities (MESP, 2011) and some 500 sub-associations with many endemic or and relic communities, which depicts the importance of the Serbian ecosystem diversity (Republic of Serbia Ministry of Environment and Spatial Planning, 2010). Serbia has harmonized the national nomenclature with the international community following the EUNIS[1] standards.

Species diversity

Species diversity in Serbia is not well researched or documented (MESP, 2011). According to a study from the MESP, Serbia hosts 39 per cent of European vascular flora, 40 per cent of Europe's bryophyte flora, 51 per cent of fish fauna, 49 per cent of reptiles and amphibians fauna, 74 per cent of bird fauna and 67 per cent of mammal species, with a territory that represents 2.1 per cent of the European continent. Serbia has numerous relic species diversity and has rich endemic species (287 Balkan endemic species or 8.06 per cent and 59 local species or 1.5 per cent) (Republic of Serbia Ministry of Environment and spatial planning, 2010).

The next table shows the number of species for several groups.

Table 9.7. Number of species registered in Serbia.

Species classification	Number of species
Flora	287 Balkan endemic and 59 local endemic
Fungi	More than 600 species of Macromycetes, from which 60 are endangered to various degrees
Fish	110
Amphibians and reptiles	44
Birds	345
Terrestrial mammals	94

Source: SEPA (2007).

Human driven pressures such as urbanization, development of agriculture, industry, mining and transport infrastructure have brought the degradation of natural ecosystems through cultivated agro ecosystems, fragmentation of habitats, overexploitation of generic and biological resources, introduction of alien species, contamination, increased level of ionizing and nonionizing radiation, and climate change. These factors altogether have resulted in a significant reduction of biodiversity in Serbia (SEPA, 2007).

The Red List of Threatened Species is the most comprehensive inventory in the world of plant and animals and assesses the risk of extinction of species and subspecies. The 1st

[1] EUNIS data on Biological Diversity (species, habitat types and sites) is collected and maintained by the European Topic Centre for the European Environmental Agency and the EIONET, and used for environmental reporting and assistance in the NATURA2000 process and coordinated to the EMERALD Network of the Berne Convention (MESP, 2011).

Red Book developed by the Republic of Serbia in 1999 contains 171 plant taxa (species and sub-species), which represent approximately 5 per cent of total Serbian flora. The 2nd Red Book contains 57 butterflies, accounting for 34 per cent of butterfly fauna in Serbia.

Genetic diversity

Serbia is rich in genetic resources with numerous varieties, sorts and types of autochthonous populations of all plants, animals and microorganisms. Serbia's potential for genetic diversity is mainly noted in forests and natural grasslands.

With regard to agriculture biodiversity, there are over 1,200 varieties of agricultural plants (of which over 740 are cereals). Domesticated animals lost genetic potential due to the enhanced production of certain sorts. More targeted action is suggested in order to maintain not only certain species but to increase their number too (Republic of Serbia Ministry of Environment and Spatial Planning, 2010).

9.6.2 Influence of agriculture on biodiversity

The major threats to biodiversity in Serbia are air, water and soil pollution. Pollution originated in the industrial sector has decreased in the last years because of the reduction of industrial activities. With farmland covering 70 per cent of total surface area of Serbia, agriculture runoff is considered a major pollutant factor. Use of pesticides and fertilizers is one of the several sources of pollution, responsible not only for soil pollution but also for the air pollution. In agricultural regions air pollution is mainly coming from the volatile toxic compounds ammonia, nitrates, phosphorus, etc. Another factor is the overgrazing in some areas, which contributes to the loss of soil resources and loss of biodiversity. While traditional harvesting and grazing practices in Serbia represent an opportunity to support rural living within the limits of biological resources, Serbia's development as a market economy and the associated economic growth brings new environmental pressures, and therefore new threats as well as new opportunities for the protection of biodiversity. Summarizing, the main threats coming from agriculture are natural habitat destruction, over-exploitation of natural resources and the effects of the use of agrochemicals and agro-machinery (Republic of Serbia Ministry of Environment and Spatial Planning, 2010). Currently, however, over-exploitation of natural resources and/or high use of agro-chemicals is rare in Serbia.

High Nature Value Farming

The concept of High Nature Value (HNV) Farming is related to the fact that certain types of farming are extremely valuable for wildlife and for maintaining biodiversity. These types of farming are characterized by low intensity land use (extensive use of machinery, fertilizers and pesticides and low densities of grazing livestock), presence of semi-natural vegetation and diversity of land cover and land use. Not all characteristics need to be present within a farming system to be considered HNV. Partly funded by the Ministry of Agriculture, Nature and Food Quality of the Netherlands, the IUCN (International Union for Conservation of Nature and Natural Resources) carried out and published in 2010 a study on Developing a National Agri-Environment Programme for Serbia (Cooper *et al.*, 2010). In this study several areas in Serbia are identified as potential HNV farming areas. The study concludes that adequate targeted support would contribute to the economic viability of HNV farming and to a modern quality of life for rural communities based on low intensity farming methods, quality food products, traditional crafts and rural tourism.

9.6.3 Institutional and policy framework for biodiversity conservation

Institutions

The Ministry of Environment and Spatial Planning (MESP) has the overall responsibility in the field of nature protection and preservation of biodiversity. The MESP is also responsible for the harmonization and implementation of environmental legislation in the Republic of Serbia and currently focuses on the EU approximation. The main competencies of the MESP are to take care of protection of nature, natural resources and biodiversity, to designate protected natural resources under its competences, to participate in the creation of management plans for protected natural resources under its competences and to provide support to local government units with regard nature protection at their districts.

In March 2002 the Provincial Secretariat for Environment Protection and Sustainable Development was formed by the Decision of Provincial Administration (Official Gazette of AP Vojvodina, number 21/2002) as an independent administrative body at the level of Province. Within the Secretariat the nature protection activities are performed by the Department of Protection and Improvement of Natural Resources and Biological Diversity, with tasks in the field of protection of nature and protected natural resources, and the preservation of ecosystems and species diversity in AP Vojvodina.

The Institute of Natural Protection of the Republic of Serbia and the Provincial Institute for Nature Conservation in the AP Vojvodina play an important role in the protection of natural resources.

Besides the MESP and the Institute for Natural Protection, certain competences in the field of nature protection are also found within the following institutions: Ministry of Agriculture, Forestry and Water Management (i.e. Genetically Modified Organisms) and the Ministry of Science and Technology (i.e. research in biology, biodiversity and biotechnologies); the Serbian Environmental Protection Agency (SEPA), which was established in 2004, is responsible for data collection and reporting on the status of biodiversity. The Fund for Environmental Protection established under the MESP by the 2004 Law on Environmental Protection has the mandate to include biodiversity conservation and natural protection in its financing objectives.

Legal and policy framework

The protection of nature in the Republic of Serbia is regulated by the Law on Nature Protection of 2009 (Official Gazette of RS, No. 36/09) and the Law on Amendments to the Law on Nature Protection of 2010. The aim of this legislation is the protection and conservation of natural, biological, geological and landscape diversity, as a part of the environmental responsibility of the Serbian state (GoRS, 2011). This law is aligned by Serbia with ratified international treaties in the area of protection of biodiversity (see Box 9.1) as well as with the relevant EU Council directives on the conservation of natural habitats and wild fauna and flora, and on the conservation of wild birds.

The policy framework for nature conservation includes: the National Strategy for the Accession to the European Union adopted in 2005, the National Strategy for Sustainable Development adopted in 2007, the National Environmental Protection Programme 2010-

2019 adopted in 2010, the Spatial Plan of the Republic of Serbia of 1996 (later replaced by the Spatial Development Strategy of Republic of Serbia), the Forestry Development Strategy, the National Strategy for Sustainable Use of Natural Resources (in preparation) and the Biodiversity Strategy and Action Plan, which has been adopted in February 2011. The country's currently protected areas are presented in Box 9.2.

Box 9.1. International conventions related to biodiversity protection ratified by the Republic of Serbia.

- Convention Concerning the Protection of the World Cultural and Natural Heritage - World Heritage Convention (Official Gazette of SFRY - International Treaties, No. 8/74);
- Convention on Wetlands of International Importance, Especially as Waterfowl Habitat - Ramsar Convention (Official Gazette of SFRY - International Treaties, No. 9/77);
- Convention on Biological Diversity (Official Gazette of FRY - International Treaties, No. 11/2001);
- Convention on the Conservation of Migratory Species of Wild Animals (Official Gazette of RS - International Treaties, No. 102/07);
- Convention on the Conservation of European Wildlife and Natural Habitat (Official Gazette of RS - International Treaties, No. 102/07);
- Framework Convention on the Protection and Sustainable Development of the Carpathians - Carpathian Convention (Official Gazette of RS - International Treaties, No. 102/07);
- UN Convention to Combat Desertification in Countries Experiencing Serious Drought and/or Desertification, Particularly in Africa - UNCCD (Official Gazette of RS - International Treaties, No. 102/07);
- UN Framework Convention on Climate Change and Kyoto Protocol (UNFCCC);
- Convention on International Trade in Endangered Species of Wild Fauna and Flora - CITES (Official Gazette of FRY - International Treaties, No. 11/2001).

Source: GoRS (2011).

Box 9.2. Protected areas for nature conservation in the Republic of Serbia.

Protected areas in Serbia account for 517,918 ha or 5.9% of its total territory, divided in:

- 5 national parks (158,986 ha)
- 6 nature parks (213,303 ha)
- 16 landscapes of outstanding features (45,656 ha)
- 67 nature reserves (88,868 ha)
- 42 cultural and historical areas (2,507 ha)
- 315 natural monuments (8,598 ha)

For comparison: at the EU-27 level the protected areas for biodiversity in percentage of total territory added up to 14% in 2010. In Serbia the objectives of the MESP is to achieve 10% of total territory as a protected area by 2015 and 12% by 2021. Natura 2000 areas are expected to be implemented in Serbia no later than by its accession to the EU.

Source: GoRS (2011).

Nature conservation instruments and monitoring systems of biodiversity

The Law on Nature Protection establishes the protection regimes and establishes the obligation to approve and adopt management plans for a 10 year period, describing the method for the implementation of protection, use and management of protected areas, protection guidelines and priorities, and the conservation of the natural values of protected areas. The manager of the protected area is responsible for safeguarding internal order and protection. The law defines the system of permits issued by the manager and the use of biological, biotechnical and chemical agents in protected areas.

There is no centralized database or coordinated monitoring of biodiversity at the national level. The level of biodiversity monitoring is incomplete and fragmented. The quality and quantity of data is heterogeneous, not standardized and frequently not comparable with data in other European countries.

Despite the efforts and the commitment shown by the Republic of Serbia to the conservation of biodiversity and natural areas through the ratification of international agreements, the adoption of a national legal framework and the establishment of a large number of protected areas, Serbia nevertheless encounters difficulties in implementing policies and strategies. The reason might be found in the lack of financial resources and lack of appropriate institutional structures for biodiversity conservation, and the lack of monitoring and information systems.

9.7 Summary and conclusions

Summary

Agriculture makes use of natural resources like water and land. Degradation (e.g. water and soil pollution) or exhaustion (e.g. overgrazing) of these resources and the related biodiversity loss are examples of the negative externalities of agriculture production, while traditional harvesting and grazing practices in Serbia represent an opportunity to support rural living within the limits of the natural resources available.

Approximately 2 per cent of total fresh water abstracted in Serbia in 2009 was used for agriculture, inter alia through irrigation. The main agricultural sources of water pollution are the uncontrolled effluents of livestock farms (e.g. animal manure), crop and slaughterhouse wastes, chemical fertilizers and pesticides. Fertilizer and pesticide use in Serbia is, however, relatively limited compared to most EU countries. Eutrophication is the process by which a body of water acquires a high concentration of nutrients, especially phosphates and nitrates. There are indications that the concentration of nutrients in the rivers of Serbia were higher in the 1980s than in the 1990s and later years. However, estimating the water polluting effects of agricultural practices is hampered by the lack of means for monitoring ground water quality.

Agriculture inevitably lowers soil fertility by removing soil nutrients incorporated in the harvested crops. Excessive cropping or grazing can reduce soil-nutrient levels and degrade the soil structure. Poor drainage has led to waterlogging, salinization and erosion. Soil salinization is evident in Vojvodina, more than in any other part of the country. Exposure of soils to wind and rain during cultivation contributes to the erosion of the fertile surface. The lack of systematic soil monitoring and analysis hampers developing appropriate insights into the country's soil status, and consequently a strategy of what to do to improve soil fertility.

Serbia has a potential for the growth of the organic farming sector, currently export driven, but mainly in rural areas where conventional production is not possible or where sources of pollution are not nearby. Furthermore, during the past fifteen years Serbian farmers could not afford synthetic inputs, which has resulted in low levels of prohibited residues and implies that in Serbia conversion periods are shorter than in the most developed countries. However, the potential for growth of this sector is limited by the reduced agriculture budget for organic production, capital constraints and a poor supply chain organization, which leads to inefficient production, processing and marketing. Suggestions to improve the organic farming sector focus on credit facilities, education, promoting use of modern technology and propagation material, and support for associations and cooperatives.

Serbia's territory is very rich in biodiversity. The main threats coming from agriculture are natural habitat destruction, over-exploitation of natural resourcres and the effects of overuse of agrochemicals and agro-machinery, although the latter is rare in Serbia.

Conclusions

The analysis of environmental impacts of agriculture in Serbia encounters two important limitations. First, there is a lack of data or data available only over a short period. Second, the policy and institutional framework has been established only very recently or is still in development, which makes a satisfactory policy impact assessment impracticable.

Serbia has demonstrated its interest and commitment to protect the environment and the country's biodiversity through the ratification of international agreements and the adoption of a national legal framework. However, some EU directives in the field of environment (e.g. Nitrates Directive) have not been implemented yet. In the programme for financial support to agricultural production in Serbia, cross-compliance measures with environmental, animal welfare, public, animal or plant health standards do not exist. Serbia is facing difficulties in implementing policies and strategies for adequate environmental protection. Besides a lack of financial resources, the national legal framework is of (very) recent date and lacks appropriate institutional structures for implementation and enforcement of regulations. Central collection of data and monitoring of environmental indicators are missing or are fragmented among diverse projects. Finally, investments in agriculture for irrigation systems, drainage and flood protection, or for waste management systems (in slaughterhouses) would contribute much to reaching environmental protection targets. However, the economic instruments in policies to encourage farmers and other actors in the agri-food supply chain to be environmentally responsible may be too weak at present to get investment in environmental management to start up.

Literature and websites

Antić-Mladenović, S. and V. Ličina, *Erosion in Serbia-State, legislative, problems, tasks.* Faculty of Agriculture - University of Belgrade, Belgrade, 2010.

Bogdanov, N., *Mainstreaming environment into forestry and agriculture policies and operations.* Belgrade, 2007.

Cooper, T., Pezold, T., (eds.), C. Keenleyside, S. Đorđević-Milošević, K, Hart, S. Ivanov, M. Redman and D. Vidojević, *Developing a national agri-environment for Serbia.* IUCN Programme Office for South-Eastern Europe. Gland, Switzerland and Belgrade, 2010.

EC, *An analysis of the EU organic sector.* DG for Agriculture and Rural Development. Brussels. 2010. Available at: http://ec.europa.eu/agriculture/analysis/markets/organic_2010_en.pdf.

EEA and WHO, *Water and health in Europe: a joint report from the European Environment Agency and the WHO Regional Office for Europe.* Copenhagen, WHO Regional Publications, Copenhagen, 2002.

Environmental Reporting, *Fresh water-state and impacts.* SOER, Serbia, 2010. Available at: www.report.sepa.gov.rs/soer-2010-serbia/commonality-freshwater/soer-2010-serbia-3.

European Commission, *Organic Farming Guide to Community Rules.* European Commission DG for Agriculture, Brussels, 2001.

European Commission, *Agricultural and Rural Development.* 2008. Available at: http://ec.europa.eu/<agriculture/rurdev/index_en.htm. Accessed 31 March, 2011.

European Environment Agency, *Fresh water Serbia, the state and impacts 2010.* 2011. Available at: www.eea.europa.eu/soer/countries/rs/soertopic_view?topic=freshwater.

Eurostat, *Eurostat Statistics.* Brussels, Luxembourg. Available at: epp.eurostat.ec.europa.eu/portal/page/portal/statistics/search_database. Accessed 23 May 2011.

FAOSTAT, *FAOSTAT.* Rome, FAO. http://faostat.fao.org. Accessed 24 May 2011.

Gaston, K.J and J.I. Spicer, *Biodiversity: an introduction. 2nd Edition.* Blackwell, New York, 2004.

GIZ, *Organic Agriculture in Serbia.* Available at: http://www.organic-market.info/easyCMS/FileManager/Dateien/GIZ_Organic_Agriculture_in_Serbia.PDF. Deutsche Gesellschaft fuer International Zusammenarbeit (GIZ) GmbH. Economic Development Program ACCESS, 2011.

GoRS, *Answers to the European Commission' Questionnaire Chapter 11: Agriculture and Rural Development.* 2011.

GoRS, *Answers to the European Commission' Questionnaire, Chapter 27: Environment.* 2011.

FOAM, *Growing Organic.* Available at: www.ifoam.org/growing_organic/2_policy/case_studies/serbia_cases_studies.php.

MAFWM, *Serbia Agriculture 2011.* Serbian Ministry of Agriculture, Forestry and Water Management, Belgrade, 2011. Available at: www.mpt.gov.rs/download/OtkrijteSrbijuanet.pdf.

Maslac, T., *Organic Agriculture in Serbia.* USDA Foreign Agricultural Service, Belgrade, 2009. Available at: www.gain.fas.usda.gov/RecentGAINPublications

Maslac, T., *Serbia adopts new set of agriculture laws.* USDA Foreign Agricultural Service, Belgrade, 2010. Available at: www.gain.fas.usda.gov/RecentGAINPublications.

MESP, *Biodiversity Strategy of the Republic of Serbia for the period 2011-2018.* Ministry of Environment and Spatial Planning, Belgrade, 2011.

Republic Hydro Meteorological Service of Serbia, (n.d.)., *Republic Hydro meteorological Service of Serbia.* Available at: http://www.hidmet.gov.rs/eng/kvalitetvode/index.php.

Republic of Serbia, Ministry of Environment and Spatial Planning, *First National Report of the Republic of Serbia to the United Nations Convention on biological diversity.* Belgrade, 2010.

SEPA, *Environment in Serbia-an indicator based review.* Serbian Environmental Protection Agency, Belgrade, 2007.

SEPA, *Environmental Indicators in Serbia: A brief guide.* UNDP, Serbian Environmental Protection Agency, Belgrade, 2007.

Simić, D., *Environmental Policy in South-Eastern Europe.* UNDP, Belgrade, 2007.

SORS, *Protection against damaging water effects in the Republic of Serbia, 2010.* Statistical Office of the Republic of Serbia, Belgrade, 2011. Available at: http://webrzs.stat.gov.rs/WebSite/repository/documents/00/00/32/72/ZS10_125_eng.pdf.

SORS, *Wastewater discharged in the Republic of Serbia, 2010.* Statistical Office of the Republic of Serbia, Belgrade, 2011. Available at: http://webrzs.stat.gov.rs/WebSite/repository/documents..

SORS, Irrigation in Republic of Serbia. Available at: http://webrzs.stat.gov.rs/WebSite/repository/documents/00/00/30/80/ZS20122010eng.pdf.

UNFCCC, *GHG emission profiles for non-Annex I Parties.* 2011. Available at: http://unfccc.int/ghg_data/ghg_data_unfccc/ghg_profiles/items/4626.php.

USDA, *Organic Agriculture in Serbia.* Belgrade: USDA Foreign Agricultural Services, Belgrade, 2009. Available at: www.gain.fas.usda.gov/RecentGAINPublications.

World Bank, *Integrating Environment into Agriculture and Forestry - Serbia - Country Review.* World Bank, Washington DC, 2007.

Chapter 10

10 Animal and plant health in Serbia

10.1 Animal health

10.1.1 The animal sector

Serbia's livestock sector utilizes about 1.7 million hectares. The number of grazing animals has been constantly decreasing over the last two decades, with the sheep flock mostly affected in the mid 1990's whereas the decline in the number of cattle and pigs has largely taken place during the period 2000-2010. In 2010 the number of cattle, pigs, poultry and sheep and goats was about 940,000, 3.5 million, 20.2 million and 1.5 million, respectively (SORS, 2011). Most livestock production in Serbia takes place in small-scale, family operated farms (see also Chapter 4).

Processing units of animal products (milk and meat) are typically of a small-scale nature as well (see Chapter 6). Most processing units therefore may lack the financial means to invest in hygienic conditions that meet EU standards. Indeed, the number of dairy plants and slaughterhouses that are allowed to export to the EU are rather limited: only 2 dairy firms and 5 slaughtering houses in 2010 (GoRS, Chapter 12: 214). Legislative arrangements have been made in Serbia in order to comply with EU requirements on sanitary conditions (see Chapter 12 in GoRS). However, what is needed for the improvement of the overall sanitary conditions is an investment programme of businesses that want to respond to the opportunities the EU market is offering and well-trained personnel that can implement the legislation. Next to that the number of skilled and knowledgeable meat inspectors needs to be sufficient and inspection procedures should be updated to the modern norms. This points at a challenge at the institutional level, which relates to the efficiency with which the veterinary services are organized and to the availability of resources to pursue effective control and eradication.

Due to its small-scale farming structure, a large part of milk, meat and egg production is not offered to processors, and is used for own consumption and/or sold on open farmers' markets (see Chapter 6). It is clear that part of the annual (semi-commercial and subsistence) production bypasses conventional recording and monitoring systems. This suggests that veterinary inspection and monitoring of animals at this point in the supply chain is likely to be inadequate and not comprehensive.

Statistics on notifiable animal diseases (outbreaks, infected cases, control measures adopted and so on) in Serbia are available from the World Animal Health Organization (OIE). The following sections rely heavily on that source. However, it is stressed that these figures cover reported cases only. Inevitably, accurate reporting is more difficult the greater the share of production that has no contact with the veterinary services and falls outside the commercial marketing channels.

The following sections cover the most infectious and economically significant animal diseases (List A diseases), less infectious diseases that have also serious economic consequences (List B diseases) and diseases that can be passed from animals to humans (zoonoses), followed by a discussion of the legislative framework and the resources available for adequate animal health control. The annex contains a list of reported diseases and the most recent year of their occurrence in Serbia.

10.1.2 List A diseases

The OIE (Office International des Epizooties)[1] classifies 15 animal diseases as List A diseases. Eight of these diseases have never been reported in Serbia. Of the remaining seven, Serbia has official FMD-free status and rinderpest-free status since 2006, while sheep and goat pox have not occurred since 1955. In contrast, blue tongue (last occurrence in 2002), highly pathogenic avian influenza (in 2006), Newcastle disease (in 2007) and classical swine fever (in 2010) all have occurred in the last decade. The EU has disease-free status for most of these diseases.

Classical swine fever (CSF)

Classical swine fever has occurred in several spots in Serbia in recent years, with outbreaks in October 2007 and November 2010. The latest outbreak in 2010 was in one breeding herd in Vojvodina, where 201 pigs were affected, followed by one case a few days later on a farm in a neighbouring village. To prevent further spreading of the disease, over 8,750 animals were destroyed. Generally, control measures in use are surveillance, monitoring, vaccination and zoning. Stamping out is used only occasionally. Classical swine fever is prevented and suppressed by constant application of zoo-sanitary and bio-safety measures, in addition to constantly maintaining immunity against the classical swine fever virus. All holdings that keep and raise pigs during the whole year are obliged to carry out vaccination with attenuated CSF vaccine. The government's veterinary services conduct that work and keep records of the vaccination status of farms.

Newcastle disease

Newcastle disease, a milder version of bird flu, is highly contagious among birds but poses no hazard to humans. However, humans are known to develop conjunctivitis after contact with contaminated poultry meat or eggs. In 2007 Serbian authorities reported 9 outbreaks of the Newcastle disease to the OIE. According to the early January 2007 OIE report, more than 5,000 affected poultry of various kinds died of Newcastle disease or were destroyed in central and northern Serbian villages. As a response, Serbia's Veterinary Directorate established protection zones and surveillance zones around the outbreak villages. Intensive vaccination began with the financial support from the Serbian government, yet that could not prevent further outbreaks in the country later that year.

1 See www.oie.int for an up-to-date country diseases status, based on the WAHIS (The World Animal Health Information System) an internet-based computer system that processes data on animal diseases in real-time and then informs the international community.

Avian influenza

The last occurrence of avian influenza in Serbia in March-April 2006 consisted of only a few cases of infected birds. The occurrence was quickly contained without spreading to humans. Serbian officials have been undertaking extensive planning for any future outbreaks, such as efforts to implement a National Action Plan for the control of possible avian influenza outbreaks.

10.1.3 List B diseases

The World Animal Health Organization's List B contains over 90 transmissible diseases that are considered to be of lower socio-economic and/or public health threat than those of List A, largely because of a slower rate of spread, although these diseases may still be significant in the international trade of animals and animal products. This section reports on those with most relevance to Serbia, which are anthrax, bovine tuberculosis, brucellosis, rabies and BSE.

Anthrax bacteria spores lie dormant in the soil and can become active under conditions such as heavy rainfall, flooding or drought. Animals that are not vaccinated often die quickly after being exposed to the bacteria. Anthrax has occurred in Serbia 149 times since 2007 to mid-2011, with ten animals that have died. An outbreak of anthrax has been reported in Serbia in June 2011. The outbreak caused four deaths of animals. Authorities launched vaccination of cattle in two villages and banned the use of milk and meat from the animals in the anthrax-infected zone. Furthermore, quarantine arrangements and movement controls were instituted to prevent further spread of the disease.

Bovine TB outbreaks occur 30-40 times a year. Although the epidemiological situation considering BTB in cattle is favourable in the country as a whole, there is a persistent TB problem in some endemic areas that proves to be difficult to eliminate. In evaluating outbreaks in Vojvodina (South Bačka region) between 2004 and 2007, Pusic *et al.* (2008) point at the difficulty in tracing and controlling BTB, which is due to a lack of diagnostic and control programs in the years that preceded, illegal cattle movements and trading of unmarked infected cattle. The persistence of infection is mostly linked to the traditional extensive breeding system and free-ranging cattle. In these areas cattle herds are managed in semi-free ranging system, which requires a more specialized style of management for gathering of animals. This is often very difficult to obtain and some animals or even whole herds of smaller size eventually escape testing schemes. The untested but infected animals create a reservoir of infection. An extensive use of outlying, humid and poor quality pastures facilitates cattle to cattle spread.

Rabies is common in Serbia, as in its neighbouring countries. The average number of the cases of rabies is 80-200 annually. The main rabies source in the wild is red fox. The greatest number of rabies cases within domestic animals are recorded in the population of dogs and cats, especially in rural areas. Cases of cattle, equine, sheep and goat rabies are individual. Registration and marking of dogs and cats is obligatory, as well as vaccination which must be conducted once a year. The first general oral vaccination of wild animals in the entire country was undertaken in November and December of 2010, as part of the first component of the EU Instrument for Pre-Accession Assistance, as part of the support for transition and institution-building, in the project 'Technical support for control and eradication of classical swine fever and rabies in the Republic of Serbia', which started in January 2010.

BSE (Bovine Spongiform Encephalopathy, or the mad cow disease) has not been reported in Serbia either in the domestic herd or within imported cattle. According to the latest OIE evaluation (as set out in resolution 17 of the 79th General Session in May 2011), Serbia has a BSE-free status.

Current situation, prospects and conclusions

State veterinary services are the responsibility of the Veterinary Directorate within the Ministry of Trade, Agriculture, Forestry and Water management. The Directorate has a very wide remit, which includes: preparing plans for preventing and combating animal diseases; establishing and enforcing internal and border controls with penalties for transgression; preparing and disseminating guidelines for disease control and biosecurity, and overseeing their implementation; and registering and controlling facilities producing food of animal origin and feed production facilities.

Milk and meat processing units are generally in poor condition; they lack adequate sanitary and health controls, and the sanitary conditions for the transport of animals products are not up to EU standards (Twinning 2011 fiche). The capacity of the veterinary service to examine potential risks arising within the animal food and feed chain needs enhancement. Building capacity in the areas of food safety and animal welfare is presently subject to the IPA 2011 program.

The number of slaughterhouses and meat processing units approved for export to the EU are limited. In December 2009 there were 5 slaughterhouses for cattle, 4 for pigs and 4 meat processing (beef and pork) units that were approved (SANCO, 2009). In October 2010 an FVO mission came out to evaluate the operation of controls over the production of fresh meat and meat products destined for export to the EU. Four out of six establishments visited were found to be largely compliant with EU requirements, with a limited number of deficiencies. In one other establishment substantial deficiencies were found and another one was found not to be in compliance, having been closed since July 2009. The latter one has been delisted.

In early 2010 Serbia did not have an establishment authorized to export milk and milk products for human consumption to the EU. An FVO mission came to Serbia in February 2010 to evaluate the operation of public health controls at two milk processing establishments. However, while one of the two dairies complied with all necessary requirements, the competent authority could not demonstrate that all criteria for certification of heat-treated dairy products (as laid down in Council Directive 96/93/EC) from Serbia into the EU could be met.

The process of upgrading to EU standards of establishments producing food clearly emerged as a critical issue. Results of inspections, field trips carried out within EU funded projects and FVO missions suggest that a large number of establishments are not yet in compliance with EU standards and that the general situation should be tackled with the utmost priority.

Serbia is in the process of harmonizing its national legislation with the EU *acquis* on issues of food safety (see GoRS, Ch. 12, 2011). Despite impressive progress in this process, it is clear that effective implementation of the provisions, instructions and guidelines with respect to animal health and conformity with EU standards is a huge challenge. The repeated occurrence of a number of infectious animal diseases in Serbia require constant attention and alertness from the country's veterinary service. A major challenge arises from the small-scale structure of the sector, which impedes appropriate monitoring and control.

Implementation of legislative requirements also require huge investments in the agri-food businesses to improve their hygienic situation and to comply with all sanitary requirements. The implementation of HACCP (Hazard Analysis and Critical Control Points) in the sector is crucial in this respect. All animal products and animal feed related food business operators are obliged to implement a HACCP system from 1 January 2009 (for production, processing and trade of food from plant origin, HACCP implementation is mandatory for from June 2011). However, the number of slaughterhouses, dairy units, cooling stores and other relevant units that have implemented a HACCP system is still far below the total number of establishments that have been approved for production (see p. 216, comparing with p. 214 in GoRS, Chapter 12, 2011).

10.2 Plant health

10.2.1 Overview of agricultural and horticultural crops

Almost 85 per cent of Serbia's total cultivated land is used for field crops, including fodder crops like clover, lucerne and green maize (see Table 4.4 in chapter 4). The fruit culture accounts for about 8 per cent and the remaining 7-8 per cent is devoted to vegetables, viticulture and several smaller crops (e.g. tobacco, beans). The area planted to grains is over 1.9 million hectares, about 60 per cent of the total crop area (largely wheat and maize), while oilseeds accounts for 300-350,000 hectares. Sugar beets and potatoes are other major crops, yet on a 60-80,000 hectares. Fruits and vegetables cover a wide range of different products, due to its climatic conditions. Apples, plumes and a range of berry fruits dominate fruit production while tomato, pea and cabbage cover major areas under vegetables.

In what follows, information will be given about pests and diseases that occur in the arable and horticultural sectors, and the methods that were adopted by farmers to prevent or control them. Unfortunately our literature study has not uncovered any comprehensive studies on this topic for Serbia, so the following brief and selective overview is not guaranteed to reflect the relative prevalence of the disease covered, the importance of the economic losses it inflicts or the relative priority given to research and extension efforts to improve the situation.

10.2.2 Plant diseases in arable and horticultural crops

Diseases are considered to be among the most important yield-limiting factors in Serbia's cereal crops. *Blumeria graminis tritici*, the causal agent of powdery mildew is a regular and economically important disease of wheat in Serbia (Jevtic, 2009), and pathogens like rusts, common bunt and dwarf bunt are significant too. Yield losses up to 40 per cent are not exceptional, as an example from Romania illustrates (Mantaguihan *et al*, 2011). Chemical seed treatments can effectively control these diseases, yet in a typically low-input production as in Serbia, such treatment is little applied. Maize, Serbia's major cereal crop, is affected by the western corn rootworm, a beetle that migrated from the US and was identified in the early 1990s in Serbia. Rootworm may cause yield losses of 0.5-1 tonne/ha, and even bigger damage to the harvest is not uncommon (News.pu, 2002).[1] A traditional maize to soybean crop rotation is the main means of controlling rootworm infestations in most of the United States. Integrated pest management (IPM) has been introduced through a foreign donor-led programme to control the western corn rootworm (World Bank, 2007). Although focused on the rootworm, the programme also aims at helping to implement locally sustainable IPM strategies for maize in general through participatory training and research. No recent information could be found on the extent to which this programme has been rolled out in Serbia.

Economic losses are due to lower yields and lower quality. Damaged grain is only fit for use as animal feed, even though it has been grown for bread making. Potatoes are also highly sensitive to diseases that affect the product quality and their use. EU's latest FVO mission to evaluate the phytosanitary situation of Serbia's potato production took place in November-December 2009 (SANCO, 2010), in view of a possible export of ware potatoes to the EU. In particular, the situation of *Clavibacter michiganensis ssp. sepedonicus* was evaluated in view of a possible recognition of Serbia as being free from this organism. Overall, the report concludes that the plant health services in Serbia are well structured, with a clear division of tasks and competent inspectors. The national legislation for plant health and the relevant guidelines have been revised in 2009. Serbia has established a system of official surveys to determine the status of harmful organisms of concern to the EU, which indicate that the majority of these, and in particular *Clavibacter michiganensis ssp. sepedonicus*, are not present in Serbia. However, the FVO considers the surveys as too limited and targeted on only a too small proportion of the national production to be able to determine the disease status of Serbia's ware potato. For potential future export to the EU, the export procedures in place should ensure that only potatoes that comply with the EU's import requirements are certified. The FVO emphasizes that smaller production units should be adequately represented in the sampling.

There are many more pests and diseases in the arable and/or horticultural crops in Serbia. One more example refers to viticulture, where it has been estimated that phytoplasma *Flavescence dorée* (FD) endangers over 40 per cent of vineyards in Serbia, with a tendency to develop further (Sivčev *et al.*, 2010). Yet, the available literature largely exclusively points at the agronomic impacts (on yields) and technical measures to combat pests and diseases (applying plant protection products, changing farming practices, etc.), while there is little to no analysis of the economic aspects of plant health. Insights into the costs and benefits of prevention and of responses to plant health problems would, however, help to target policies and assistance programmes in order to achieve standards that comply with those in the EU.

10.2.3 Current situation, prospects and conclusions

The Ministry of Agriculture, Forestry and Water Management (MAFWM) is in charge of policy making in the phytosanitary field and of supervision over the legality and appropriateness of the work of the authority within the MAFWM. The Plant Protection Directorate, the Forestry Directorate and the General Inspectorate are the bodies in the Ministry conducting the activities within the phytosanitary field. Harmonisation of relevant

[1] Purdue Professor Responds to Accusations of U.S. Agroterrorism, www.news.uns.purdue.edu

laws was in an advanced stage in early 2011, yet not fully completed,[1] while adoption and implementation of the legislation into procedures and guidelines via rulebooks is programmed for (the rest of) 2011 and 2012 (see GoRS, Chapter 12, 2011). An important element is the monitoring and evaluation system of compliance. The setting up of controls for maximum residue levels, for example, is still under preparation, and a residue monitoring program of plant protection products in food of plant origin -- to observe the levels of residues with the aim of health risk assessment of the population of Serbia -- is planned for 2011-2013. This indicates that it will take a few years' time to get the systems of implementation, monitoring and evaluation in place with regard to plant health and phytosanitary requirements.

On the international level, Serbia is already a member of the CODEX Alimentarius, the European Plant Protection Organization (EPPO), the Convention of Biodiversity (CBD), the Union of Protection of New Varieties of Plants (UPOV), the World Intellectual Property Organization (WIPO) and the European Cooperative Program for Crop Genetic Resources Networks (ECP/GR); and it is a signatory of the Aarhus Convention and the International Plant Protection Convention (IPPC). Serbia is not yet a member of the WTO, but is moving in that direction with adoption of food safety and phytosanitary regulations. If Serbia joins the WTO, the country can take part in the international standard-setting body of plant protection, which is recognized under WTO-SPS agreement as providing the framework within which countries take action and cooperate to control pests and prevent their spread across national boundaries.

10.3 Conclusions

Much legislation and many procedures have been transposed into rules and guidelines for all participants in the agri-food chain. However, legislative and administrative changes alone do not guarantee a proper implementation of the veterinary and phytosanitary *acquis* of the EU. The implementation of regulations with the rigour and expertise necessary for them to function needs appropriate institutions and effective organizations, and the time to build up the expertise in this area. Twinning projects focusing on capacity improvements in these areas have been started only recently. Yet, the complexities are high, while inspectors, extension officers and farmers all need to be trained; building up the capacities to implement the regulations on animal and plant sanitary control just takes time.

The small-scale and fragmented structure of Serbia's agriculture -- at the primary, processing and trade levels -- makes monitoring and control on animal and plant health diseases difficult. Furthermore, investments in production facilities and methods that help to achieve higher standards fall short of what is necessary in order to comply with EU requirements.

There are, however, important differences between the plant and animal sectors. First, a substantial share of the plant sector is internationally competitive, and its output is actively traded. The phytosanitary standards reached by these products are clearly conformable with those required by international trade. Moreover, any plant disease outbreaks or pest infestations that might occur in the export sector will usually result in

[1] For instance, the law on genetically modified organisms, in force since June 2009, does not conform to the EU legislation and needs amendment to comply with EU requirements in this field.

nothing more serious than local production losses, with particular consignments failing to get a Plant Health Certificate from the Serbian quarantine authorities and/or particular shipments of goods being refused entry at export destinations. In contrast, given the highly infectious nature of List A and some List B diseases (a number of which remain endemic in Serbia), the potential trade consequences of an outbreak of one of these diseases are catastrophic, involving a ban on all imports from the source country or region until the disease has been eradicated according to the protocols agreed within the OIE, and international confidence has been restored.

Currently, Serbia's animal sector is only very little involved in international trade, and it is envisaged that animal products would not be competitive on the EU market (see Chapter 8). Therefore, the likelihood of Serbia becoming a significant exporter of animal products is small. Still, even with small export flows to the EU, the country's health status should be indisputable and up to standards with EU requirements in order to maintain confidence of trading partners in EU's animal products. Serbia's animal sector does not have that status yet.

Literature

GoRS, *Answers to the European Commission's Questionnaire, Chapter 12: Food safety, veterinary and phytosanitary policy*. 2011.

Jevtic, R., *Virulence of* Blumeria graminis tritici *in Serbia (2000-2009).* Poster presentation at the 12th International cereal rusts and powdery mildews conferences, 13-16 October, Antalya, Turkey, 2009.

Mantaguihan, J.B., K.M. Murphy and S.S. Jones, 'Control of common bunt in organic wheat.' In: *Plant disease* 95 (2011) 2, pp. 92-103.

Pusič, I., S. Savic-Jevdenic, D. Lalosevic, D. Bugarski, J. Prodanov and Ž. Grgic, An epidemiological evaluation of Bovine Turberculosis in South Backa region of Vojvodina Province. *Lucrari Stiinlifice Medicina Veterinara* Vol. XLI, 2008, Timisoara, pp. 264-270.

SANCO, *FVO report residues and contaminants (live animals and animal products)*. DG SANCO No. 2009-8194, 2009.

SANCO, *Final report of a mission carried out in Serbia from 30 November to 04 December 2009 in order to evaluate the plant health controls in the potato sector and in particular the situation of Clavibacter michiganensis SSP. Sepedonicus, DG(SANCO) 2009-8359, Ref. Ares(2010)316738 - 08/06/2010*. 2010.

Sivčev, B.V., I.L. Sivčev Z.Z. Ranković Vasić, 'Plant protection products in organic grapevine growing.' *Journal of Agricultural Science* 55 (2010) 1: pp. 103-122.

SORS, *Statistical Yearbook of the Republic of Serbia 2011 - Agriculture*. Statistical Office of the Republic of Serbia, Belgrade, 2011.

World Bank, *Integrating environment in agriculture and forestry: progress and prospect in eastern Europe and Central Asia.* World Bank, Washington DC, 2007.

Annex to Chapter 10. List of reported animal diseases and year of their last occurrence in the Republic of Serbia.

Disease	OIE code	Last occurrence
Foot and mouth disease (free status obtained in 2006)	A-010	1996
Rinderpest (free status obtained in 2006)	A-040	1883
Bluetongue	A-090	2002
Sheep pox and goat pox	A-100	1955
Classical swine fever	A-130	2010
Highly pathogenic avian influenza	A-150	2006
Newcastle disease	A-160	2007
Anthrax	B-051	2008
Aujetzsky's disease	B-052	2009
Echinococcosis/hydatidosis	B-053	2010
Leptospirosis	B-056	2011
Q fever	B-057	2010
Rabies	B-058	2011
Bovine brucellosis	B-103	2010
Trichinellosis	B-062	2011
Bovine tuberculosis	B-105	2011
Bovine cysticercosis	B-106	2010
Enzootic bovine leucosis	B-108	2011
Malignant catarrhal fever	B-114	2007
Caprine and ovine brucellosis (*Brucella melitensis*)	B-152	2011
Caprine arthritis/encephalitis	B-153	2011
Enzootic abortion of ewes (*ovine chlamydiosis*)	B-156	2007
Maedi-visna	B-161	2011
Porcine cysticercosis	B-252	2009
Porcine brucellosis (*Brucella suis*)	B-253	2006
Avian tuberculosis	B-303	2008
Fowl cholera	B-306	2010
Fowl pox	B-307	2007
Fowl typhoid	B-308	2010
Infectious bursal disease (Gumboro disease)	B-309	2009
Avian mycoplasmosis	B-311	2007
Spring viraemia of carp	B-404	2009
American foulbrood	B-452	2011
Nosemosis of bees	B-454	2011
Varoosis	B-455	2011
Toxoplasmosis	C-612	2007
Blackleg	C-614	2006
Infectious footrot in sheep	C-702	2006
Enterotoxaemia	C-704	2011
Salmonella enteritidis and *S. typhymurium* infection	X-001	2010
Swine erysipelas	X-003	2006
Listeriosis	X-004	2010
Bovine enzootic mastitis	X-007	2011

Source: MAFWM, internal document.

Chapter 11

11 Expected consequences for Serbia and the EU of Serbia accessing the EU

11.1 Introduction

This chapter draws on the previous chapters of the report in order to derive the likely consequences for Serbia's agricultural and agri-food sectors should it become an EU member in five to seven years. The assumptions about how the underlying context will have changed over this time span are set out in section 11.2. These assumptions are also used in section 11.7, which considers the implications of Serbia's entry for the current EU members.

If Serbia becomes a full member of the EU in five to seven years, this means that the country has to comply fully with the EU *acquis communautaire* by that date. Therefore, part of the discussion in this chapter focuses on the implications for Serbia of adopting the *acquis*. Logically, this discussion would be followed by a second part that explores the expected consequences for Serbia in the years following EU entry. This would, however, involve making assumptions about the precise entry terms agreed in the negotiations, as well as predictions about many exogenous factors that are hard to foresee. With almost no relevant studies to draw on, our exploration of the longer-term consequences will be limited. We will, therefore, touch on this issue only in the final part of this chapter.

Three groups of issues are explored in this chapter. The first set of issues relates to Serbia's ability to assume the obligations of the *acquis* regarding competition rules, property rights (intellectual and industrial), and free movement of agricultural products within five to seven years. The second set of issues relates to the implications of membership for agricultural markets in Serbia, and their capacity to cope with competitive pressure and market forces implied by the EU single market. The emphasis will be on how the policy changes necessary for membership affect the functioning of agricultural and food markets, against the background defined by the assumptions in section 11.2. The third group of issues concerns the budgetary implications. Since EU budget flows to Serbia imply nearly identical budget costs for the EU-27, there is a strong link in this respect.

This chapter is organized in eight sections. Following the introduction, section 11.2 presents the key assumptions underlying our future projections. Section 11.3 critically discusses the implications of EU membership for Serbian policies that relate to agricultural, food and rural development, international trade, and agri-environmental interactions. Section 11.4 explores the implications for agri-food chain developments. Implications for the institutional framework of Serbia's agriculture are discussed in section 11.5, with an emphasis on the quality and functioning of economy-wide institutions to the extent that they shape agricultural activities. Financial consequences for Serbia are explored in section 11.6. Section 11.7 takes the EU perspective and discusses the consequences of Serbian EU membership for the EU food chain and trade opportunities, and for the EU budget. Section 11.8 summarizes major conclusions and formulates a number of recommendations on how the Serbian government and the agri-food sector could take up the challenges that face them in preparing for EU membership.

11.2 Assumptions

The discussion in this chapter requires assumptions about future EU policies, and about exogenous developments in Serbia and the EU. Because part of our analysis involves the calculation of budget payments and financial flows, we make specific quantitative assumptions regarding some key variables. Assumptions are also needed regarding economic growth rates in Serbia and EU-27, the dinar-euro exchange rate, and EU policies relating to agriculture, food, trade, regional development and the agri-environment. Other aspects of our analysis are purely qualitative, and do not depend on specific numerical assumptions. It is assumed that by the time Serbia becomes a member of the EU, the EU will still consist of 27 member states.

11.2.1 Economic growth

We assume the same growth rate as the IMF (2011) short to medium term growth projections, which estimate 3 per cent growth for 2011, 5 per cent growth for 2012 and a continuing 5 per cent growth rate up to 2016. From 2004 to 2008 Serbia achieved an annual growth of over 5 per cent on average, so the projections for the years to come are close to a continuation of a trend throughout most years of the previous decade.

Annual economic growth of 2 per cent for the EU-27 as a whole is assumed. Inflation of 2 per cent for the Euro Zone is assumed, which implies that over a timeframe of five to seven years the price level will be 10-15 per cent higher than in 2011.

11.2.2 Exchange rate developments

The exchange rate between the Serbian dinar and the euro plays a key role in determining some important financial magnitudes. These include the future budget contributions of Serbia, price levels for products in Serbia relative to the EU, the estimates of budget expenditures expressed in euros and the upper limit of absorption capacities for structural funds.

We assume an appreciation of the dinar against the euro in the years up to accession. The assumption is based on the expectations that the outlook on accession supports foreign investment due to the confidence that a stable economic environment will provide for investors. Also, preparations for EU membership imply an absorption of at least a part of the informal economy, due to the implementation and enforcement of the *acquis*. The current rather significant informal economy of Serbia undervalues the market exchange rate of the dinar against the euro.

The implication of a dinar becoming stronger against the euro is that the budget contribution of Serbia to the EU in euro will be less in dinars than at today's exchange rate. However, prices relative to those in the EU will develop to the country's disadvantage as prices in dinars will imply higher prices in euro. This will make the Serbian agri-food chain less price competitive on the EU market. Moreover, when the euro depreciates against the dinar, direct payments in euro will be less in dinars. To illustrate the effects of the assumed exchange rate adjustments, we assume the market exchange rate will appreciate by 20 per cent over the next seven years up to accession.

11.2.3 EU budget

Annual EU budgets are based on the multi-annual financial framework (MFF) agreed between the European Parliament, Council and Commission in an interinstitutional agreement. The financial framework sets the maximum amount of commitment appropriations in the EU budget each year for broad policy areas (see www.ec.europa.eu\budget). The current MFF 2007-2013 will soon expire, and the discussion on the design of a new framework, and on the focus of payments and commitment appropriations, is making progress. In June 2011 the Commission presented its proposals.

The EU is modernizing its budget framework, responding to the challenges of increasing the Union's competitive position in the world. Central to the Commission's proposals are the objectives of the Europe 2020 Growth Strategy. (i.e. 'smart, sustainable and inclusive growth'). As in the wake of the economic and financial crisis most EU member states are reducing budgetary spending drastically, the EC is aware of the necessity of a responsible and most efficient use of budgetary resources, targeted at priority areas. The EC communicates its proposal as follows:

'Smart, sustainable and inclusive growth is the leading theme for this proposal. The Commission is proposing to increase the amounts allocated to research and innovation, education and SME development. It is proposing to unlock more of the potential of the Single Market by equipping it with the infrastructure it needs to function in the twenty first century. It is proposing to make the Common Agricultural Policy more resource efficient, so that it not only delivers high quality food but also helps to manage our environment and fight climate change. The theme of solidarity also runs through this proposal -- solidarity with the poorest Member States and regions by concentrating the biggest part of cohesion spending on their needs'.[1]

Amounts proposed by the Commission for the CAP add up to 371.7 billion for 2014-2020. This is the sum of 281.8 billion euro for Pillar 1 and 89.9 billion for Pillar 2 in 2011 constant figures. Compared to the 417 billion euro in the current financial perspectives, these figures imply a real reduction of the CAP budget, while the spending becomes further devaluated by inflation.[2] How this will affect the distribution of the support to the member countries is not in the Commission's proposal, but will follow in legislative proposals that were expected to be due in October 2011. Obviously, the budgetary constraints will have an impact on the EU agricultural policy in the years to come, the shape of which has been hotly debated within the Council of Ministers of Agriculture for quite some time. Furthermore, the Commission plans to allocate some 336 billion euro to Cohesion Policy for the post 2013 period, which is slightly less than the total EU budget for 2007-2013 (348 billion euro).

---

[1] Page 5 of the Communication from the Commission to the European parliament, the Council, the European Economic and Social Committee and the Committee of the Regions, A Budget for Europe 2020, Part I, COM (2011) 500 final, Brussels, 29.06.2011.{SEC(2011) 867 final} {SEC(2011) 868 final}.

[2] This includes budget (8.8 billion euro) that is required to finalize the phasing-in of direct payments in the New Member States. The Commission has reallocated certain elements of the CAP budget in order to 'free up' funds for this purpose.

11.2.4 EU agricultural policies

On 18 November 2010, the European Commission presented its Communication *The Common Agricultural Policy towards 2020* (EC, 2010) in which it lays down the outlines of the CAP after 2013. The EC wants a CAP with strong common features and wishes to maintain the two pillar structure of the policy. Yet to meet the challenge of worldwide food security, sustainable management of natural resources and mitigation of climate change, as well as the pursuit of increasing cohesion of rural areas in Europe, the Commission argues that some instruments need to change. The main issue involves the system of direct payments to farmers, which should be adapted to make them support farmers' income while linked to the provision of societal services. In addition to that basic premise, the differences in levels of payments among farmers/countries should be reduced, yet at the same time avoiding disrupting impacts for some producers and/or regions.

The Commission does not come up with a blueprint of a new CAP, but pictures three possible directions in its Communication: status quo, greening of the CAP and a radical reform. Given the discussions since the document was published and the indications in the MFF, the most likely policy direction is the greening scenario. In this policy option, the direct payments would be made contingent on adherence to a suite of greening measures that go beyond the cross-compliance requirements of the current legislation. There are some indications that direct payments are probably to be replaced by a basic payment with an extra allowance for green services. The scenario proposed would result in a result in a redistribution of direct payments implying a more equally divided support (as payment per hectare). Yet, the details on the conditions for green payments and the level of the basic payment are only highly indicative at the time of writing. For the purpose of analysing the budgetary impact of Serbia entering the EU, we therefore assume that nothing changes fundamentally in the criteria for being eligible for the CAP direct payments.

According to the MFF proposal, the present structure of the Pillar 2 on rural development policy is to be maintained. The funding will be subject to the Europe 2020 performance-based conditionality provisions, targeted to employment, innovation, climate change/energy, education and poverty reduction. Again, it is not clear how the change in conditionality of the funding will impact the financial assistance to countries with regard to rural development policies. In our budget calculations for Serbia's accession to the EU we assume that the conditions to allocate rural development support to a country in the 2007-2013 budget will also apply at the time of Serbia's accession.

11.2.5 Structural and cohesion funds

As said above, the use of funds under the Cohesion Fund will be related to the Europe 2020 strategy of growth, and will be reduced in real terms. The MFF proposal indicates that there will be increased spending on the poorest regions, redistributing budget from a group of 'transition' regions with GDP between 75-90 per cent of the EU average, to the poorest areas. We assume that the Commission's proposal for structural and cohesion funds for 2014-2020 is adopted, effectively implying that the rules for allocating these funds to countries with GDP below 75 per cent will remain unchanged.

11.3 Policy implications

11.3.1 Implications for agricultural, food, rural and structural policies

Agricultural policy

Potential effects of integration on the agricultural policy of Serbia have not yet been comprehensively examined by applying econometric or simulation models. Therefore, the policy implications will be discussed by comparing the current with the desired situation, and by referring to the experiences of countries of the last round of EU enlargement.[1] Here we assume that by 2020 CAP will not be fundamentally changed and that the principles that were applied in the previous enlargement will remain in force.

The fundamental principles of Serbia's agricultural policy have been laid down in the Agriculture and Rural Development Strategy for 2010-2020, as well as in the Law on Agriculture. The key aim of this Strategy document is to prepare the agricultural sector for EU integration. In the field of agricultural policy, significant reforms will be conducted in the coming years that will result in changes in volume of support, types of users, types of incentives and implementation mechanisms.

The overall effect of integration will depend on how effective agricultural policy will be in addressing the main challenges: 1. Consolidation of the existing farm structure and the reduction of the number of semi-subsistence farms (which is directly linked to the restructuring of other economic sectors, growth of employment and vitalization of the land market); 2. technical and technological modernization in almost all sectors of agriculture (excluding oilseed production and one part of fruit-wine production); and 3. strengthening the position of commercially oriented producers by horizontal and vertical integration.

Serbian agricultural policy already incorporates a direct producer's support system with a flat rate payment per hectare, which is (currently) limited to 100 ha per farm. Currently, about 10-15 per cent of farms benefit from direct payments. Therefore, the amount of support per hectare of Utilized Agricultural Area (UAA) in Serbia is considerably lower than the EU average, and lower than the average EU-12 too (member states since 2004). On the other hand, the experience of NMS and candidate countries of the Western Balkans suggests that budgetary support to agriculture has grown significantly in the pre-accession period (after acquiring candidate status), in order to increase the level of production and raise the sector's competitiveness prior to the negotiations process. Such a scenario can be expected also in Serbia.

The relatively modest budgetary support and the low prices of agricultural commodities (close to the lowest in the EU countries) lead to the conclusion that, in order to align with EU levels, a gradual adjustment of direct income support and prices will have to be established as has been done in the EU-12. However, under the assumption of the dinar appreciating against the euro (see section 11.2), the current gap in Serbian versus EU prices as well as in direct income payments will diminish over time, just because of this

---

[1] Some authors now believe that the differences in the effects of the transition in EU-12 caused their agricultural sectors are much more dissimilar now than was the case before the 1990s. Different models of transition, i.e. different land policies and land market and privatization processes, are key factors that contributed to a very heterogeneous agri-food sector nowadays in EU-12 .

reason. However, price developments depend on more factors than only exchange rate developments, while direct payments are subject to a phasing-in process (see also section 11.6). Hence, there are not enough elements for an objective assessment of this process of adjustment and its effects on production, trade and income in Serbia.

Recognizing the complexity of administrating direct payments in the EU-12, acceding countries were offered the choice to apply a simplified system of direct payments, the Single Area Payment Scheme (SAPS). Under the SAPS, farmers in the NMS receive a flat-rate payment per-hectare irrespective of what is produced, as long as their land is kept in good agricultural condition. Cross compliance measures do not exist in Serbia, although there is an attempt to create a system, which will form the basis for future implementation of cross compliance measures (it is incorporated in the rules regulating the status of an active holding). At this moment, in contrast to the EU practice, a precondition for the direct payments (for farms over 10 ha) is the receipt of the purchase of inputs instead of meeting standards of quality, environmental protection and/or animal welfare. Such payments are at best only partially decoupled and, hence, still affect production. Most such interventions lack any market failure justification.

In addition, all new member countries had to choose the minimum land size that could be considered entitled for a direct payment. Although the EU proposed 0.3 ha, all NMS chose 1 ha as a threshold. While this approach gives more money to larger farms, it does not make available income supplements to the numerous small farms, which are struggling to survive. In the case of Serbia, a threshold of 0.5 ha would exclude 111.5 thousand farms (according to Census 2002) from direct income support, while setting the limit on 1 ha would double this number (214.4 thousand). At present these small farms do not receive budgetary support, and EU accession is not expected to benefit them much. The problem of small farms is a delicate one as it was in the NMS, and probably causes a deepening of regional disparities between northern and southern parts of Serbia.

More difficulties will be experienced in the animal sector, which at this moment is both declining and inefficient. All incentives, including the milk premium in recent years, were constantly changed with regard to both the types of users and to the amount of support. On the other hand, land used for livestock production (meadows and pastures) in Serbia was excluded from the system of direct payments per ha. All this suggests the conclusion that no direct income compensation based on the number of cattle will be offered that would fit into the flat rate per hectare and that would open up demands for substantial compensation.

Food policy

Serbia adopted the new Law on Food Safety ('Official Gazette RS' 41/09) in May 2009 and the Codex Alimentarius. The adoption of the Law was necessary for the purpose of harmonization with EU regulations, in particular Directive 178/2002/EC, and for compliance with the Agreement on Application of Sanitary and Phytosanitary Measures of the WTO. Although WTO membership will open new opportunities for trade in agri-food products, weaknesses in national food control systems might limit this potential. The main problems in the implementation of the Law is in the part about sharing out of competences, in particular because the agricultural inspection system does not have adequate capacities to carry out the food safety controls of non-animal food products in production, processing, wholesale and exports, along with the introduction of internal controls, since these responsibilities would place an extra financial burden on its operations.

However, these adjustments require changes in the relevant institutions and considerable investments because many quality improvements demand new technologies and modern, upgraded facilities. The required changes could be facilitated by more foreign direct investment in the food and retail sector, but also by well-developed national programmes. Well targeted support will also be needed for the agri-food sector to adjust to progressively more demanding food safety standards. This assistance should come in the form of competitive grants within the framework of rural development programs (IPARD program). The experience of the EU-12 shows that only large companies could make the necessary upgrades, even with EU assistance programmes providing financial support. Many small businesses in the EU-12 had to close, either because they were too small to assume the financial burden of adjustment or because their processing facilities were too outdated.

In EC's Progress Report for 2010 emphasis is placed on the lack of progress in the field of food safety that requires substantial legislation to be enacted and enforced. Not only the food industry, but also the Ministry needs to improve capacities to take on its new competences with regard to food safety. Therefore, improving the food safety standards will be a particularly delicate issue for SMEs and those enterprises/ producers who survive at the moment due to lower prices achieved partially from the disrespect of standards. The Serbian Government can negotiate special provisions with the EU that allow low-capacity firms to continue producing for the domestic market without fully complying with EU requirements (as it was the case in Slovenia).

Rural policy

Taking into account the size of rural areas, the high proportion of the population living in them, and high relevance of agriculture for the national economy, it is obvious that rural development policy is a vitally important policy area for Serbia. Rural areas of Serbia face numerous specific socio-economic challenges such as low income levels, lack of employment opportunities, deteriorating quality of life, depopulation processes, high poverty rates, etc., which can be effectively tackled with rural development policy. The rural areas in Serbia will likely move in the same direction as those in EU-12 and South-European countries. But how fast change happens depends on the effectiveness of public policies.

Reform of rural development policy in Serbia and its adaptation to the EU model has started in 2008. The experience of the EU-10 and of the candidate countries of the Western Balkans has been used in the design of rural development support and the NPRD. Bearing in mind that the dynamics of withdrawal of SAPARD funds was very slow, and that many countries have not fully benefited from ear-marked funds due to national programmes being insufficiently harmonized with the EU rules, the Serbian approach was to put efforts on harmonization of its national policy with the requirements of the IPARD. Primarily driven by European integration and preparation of the institutional and legal framework to access EU pre-accession support, support to rural development has been increasing.

Up to now Serbia's rural development measures are already broadly aligned with those of the CAP: support for farm productivity investment; measures to improve product quality; and investment in farm diversification and rural infrastructure. In addition, field-based services that ensure access to rural development support for potential beneficiaries are also established as part of the Rural Development Network (which should serve as the basis to create LAGs). On the other hand, various key problems, as well as development opportunities, are not adequately considered, such as Less Favourable Areas measures, forestry measures, and measures for semi-subsistence farming and rural poverty.

The main obstacles of Serbia's current rural policy are its lack of a fully integrated approach or programming structure, absence of vertical and horizontal policy coordination and poor inter-ministerial cooperation. A more coherent approach and better coordination between policies and available funds is needed, pulling different resources together.

11.3.2 Implications for trade policy, animal and plant health, and policies concerning environmental impacts of agriculture

Several changes to the Serbia's current trade policy and, consequently, its trade position can be expected if it becomes a full member of the EU. First, Serbia will adopt the common external tariff (CET) of the EU for every agricultural product. Depending on whether current tariff lines are higher or lower, Serbia's trade may be affected one way or the other. Currently Serbia's most significant import tariffs are those for live animals, meat and dairy products, vegetables, fruit and sugar (see Chapter 8). EU's ad valorem tariffs for sugar and beef, and also for cereals and dairy products are particularly high for those countries which do not have any preferential agreement with the EU. Serbia, however, is a net exporter of sugar, cereals and beef, hence these changes will not affect Serbian import flows. Some of Serbia's current tariff lines might be higher than those of the EU, and then the new CET will lead to cheaper imports from non-EU countries. This may result in trade creation effects. Trade relations with CEFTA countries are not expected to be impacted by Serbia's EU membership, as the current agricultural trade conditions between the partners are that all customs duties are abolished (according to the Additional protocol, signed in February 2011, see also Chapter 8).

Second, internal EU-Serbian trade barriers created by tariff measures will be abolished. In principle, this change will have a trade creation effect, encouraging mutual trade. The impact of the latter will, however, be limited. As part of the Interim Trade Agreement of the SAA, Serbia has already tariff-free access to the EU market for all agricultural products except for a few. Also, Serbian tariffs will be reduced over a transition period of 6 years to no or only small tariffs left for a limited number of products. The consequence of the ITA is that price levels in Serbia and the EU will gradually move towards each other, leaving price differences mainly to quality differences.

By the time Serbia becomes a member of the EU, the country is probably part of the Union's external border. Hence, Serbia must be in the position to operate the degree of controls for goods along those external borders that is required to implement all EU measures for goods that cross its frontiers. Developing the appropriate infrastructure, sufficient administrative capacity and commitment for effective border control is part of Serbia's adoption of the *acquis*. The present situation of both the current border control practices and institutional development as a result of adopting the *acquis* builds confidence that Serbia will meet these requirements.

The country applies a Law on Foreign Trade Transactions that is in compliance with the rules of the WTO. Also, Serbia is in the final step for WTO membership, although some work still needs to be accomplished on issues like sanitary and phytosanitary aspects. Yet the outlook of being fully integrated in the world economy through WTO membership may further encourage the Serbian authorities to comply with the institutional requirements that such a membership requires.

The sanitary and phytosanitary (SPS) regulations are an important part of the *acquis* communautaire, specifically in respect to animal and plant diseases. Chapter 10 has reported that Serbia already took over much of the legislation and procedures with regard to the guidelines for all participants in the agri-food chain. Yet, legislative and administrative changes alone do not guarantee a proper implementation of the veterinary and phytosanitary *acquis* of the EU. The implementation of regulations with the rigour and expertise necessary for them to function needs appropriate institutions and effective organizations. Of major importance is also the necessary time to build up expertise in this area. Twinning projects focusing on capacity improvements in these areas have only started recently. Yet, the complexities are high, while inspectors, extension officers and farmers all need to be trained; building up the capacities to implement the regulations on animal and plant controls just takes time.

The small-scale and fragmented structure of Serbia's agriculture -- at the primary, the processing and the trade levels -- makes monitoring and control on animal and plant health diseases difficult. A large part of the farms are (semi)subsistence, selling some of their surplus on the informal farmers' markets, where quality and food safety inspections are conducted only on a limited scale. Furthermore, investments in production facilities and methods that help to achieve higher standards are necessary in order to comply with EU requirements. Yet the small companies lack the financial means (and access to credit) to invest in these requirements.

The big challenge for Serbia in preparing for EU accession shows up here once more: how to adopt and apply the *acquis communautaire* in a fragmented agri-food supply chain, which needs restructuring and modernisation, while at the same preventing the socio-economic decline of the sector?

Serbia is well advanced in aligning itself with the environmental *acquis*, which is part of the objective to avoid adverse impacts of agricultural activities on the environment. Once again, however, the implementation of this legislation and related measures are just in a first phase or still have to start up. Most importantly, (financial) policy incentives have to encourage and stimulate the farming community to apply the measures. Currently, however, the EU type of cross-compliance measures with environmental, animal welfare, public goods, and animal or plant health standards simply do not exist. This may be due to a lack of financial resources necessary for such measures. Nevertheless, institutional structures to implement and enforce regulations to protect the environment from adverse agricultural activities are an important part of the applying the *acquis*, and these are presently only in an initial stage of establishment in Serbia.

11.4 Implications for the agri-food chain development

11.4.1 Bottlenecks to improve competitiveness

Joining the EU's single market would subject the Serbian food supply chain to competition from mature industries and highly efficient, well-organized companies in the rest of the Union. To be ready for this competition, the Serbian agri-food chain has to tackle a number of serious bottlenecks. The structure of the Serbian farm and processing sector, and the general overcapacity of companies in the food sector, have all been described in Chapter 4 and 6, respectively. The product flows through the chain mainly pass via the open market, while a large share of production is used for household consumption. The wholesale market system seems to play a limited role as an alternative marketing channel between businesses.

Low capacity utilization, fragmentation and weak chain integration are characteristics of the Serbian agri-food supply chain, although there are a few exceptions.

Competition on the fairly saturated EU food markets is increasingly with respect to quality. In Serbia, consumer awareness of quality issues is limited to a more prosperous minority in the bigger cities. However, this group will increase in size as incomes rise and consumers become more discerning. Setting standards on food quality and food safety and making the players accept the rules of the game are important public responsibilities that are increasingly assumed by the Serbian government authorities.

As well as income growth and public policies, however, the move towards competition on quality will be pushed by the rise of supermarket chains in the country. Supermarkets pay increasing attention to quality as part of their strategy to gain market share from the traditional retail channels. If the Serbian agri-food chain wants to take part in the expected expansion of the modern retail sector, it has to match the quality of its supply with the quality demanded by the supermarkets. In setting their private standards, supermarkets normally take public standards as a minimum level. Complying with requirements set by the supermarkets -- and in their wake also the requirements set by the processing industry -- in the coming years should help the Serbian agri-food chain to prepare for possible EU membership.

Some of the (sub-)sectors in Serbian agriculture appear to be already rather competitive, seeing that they have gained a significant export flow to the EU in recent years. Examples are vegetables, sugar and some vegetable oils; but certainly the fruit sector (especially various types of berries) has achieved a strong export position in the trade relations with the EU (see Chapter 8). The latter indicates that these products do reach the quality standards necessary to penetrate the EU market.

11.4.2 Impact on the food industry

The increasing weight of large-format supermarkets in the retail sector and the changes in the procurement system will boost the trend towards further consolidation in the food industry. The major driving force is the quality issue. Supermarkets in Serbia increasingly set conditions in terms of prices and quality of the supply offered by the food industry, as well as in terms of other product attributes such as appearance, product diversity, convenience, safety, and so on. Supermarkets also demand that suppliers comply with requirements for packaging and delivering times. Processors who want to be part of the supply chain have to adjust to these demands. This requires investment in production and process technology. Much investment in modern technology needs a minimum operational size to reach the break-even point. At present, in many branches a large share of the processing companies is too fragmented to make such investments. Small processing firms will have to invest in expansion, or merge with others or form alliances to gain economies of scale, or leave the business.

Consolidation in the food industry will also be driven by supermarkets' preference for dealing with a limited number of large suppliers to minimize transaction costs. Food processors may also want to expand in order to strengthen their bargaining position vis-à-vis the large retail chains. The present low capacity usage in much of the food industry is detrimental to profitability, and will inevitably lead to a restructuring of the industry.

Foreign direct investment (FDI) should play an increasing role in the restructuring and modernization of the Serbian food industry. Basic conditions for attracting FDI are political and economic stability. As long as these two conditions are not met, foreign companies will be very cautious about investing in the country. Experience from Central and Eastern Europe indicates that the prospect of becoming an EU member can increase a country's attractiveness, even if the date of accession is somewhere in the future. Serbia's legislative framework is in place, as well as supportive policy instruments to attract foreign investors (see www.siepa.gov.rs).

11.4.3 Impact on the farm sector

The retail and food processing industry will pass on the more demanding requirements with respect to food quality and safety to the farming sector. These demands may push many small farmers out of the market when they find it hard to comply with the requirements. Small farmers often cannot make the necessary investments, because they have insufficient own resources and face problems in getting external credit. Moreover, especially in the case of unprocessed, perishable products such as fresh fruit and vegetables, large transaction costs make it more costly for retailers to deal with many small farmers rather than with a few larger suppliers. However, small and medium farmers can have a future in modern retail chains if retailers and/or food processors are willing to make investments in vertical coordination with suppliers, for example by offering farm assistance programmes providing farms access to inputs such as knowledge, techniques and credits, and enhancing their output in terms of quality and quantity.

The process of vertical contracting risks excluding small farms. However, the equity implications of such integrated chains are a justification for public policy involvement. Areas for government initiatives could include stimulating the emergence of alternative marketing structures (e.g. cooperatives), promoting associations of (small) farms to increase their bargaining power vis-à-vis the agribusiness companies and reducing transaction costs for companies dealing with small farms.

11.5 Implications for institutions and their functioning

If Serbia becomes a member of the EU, the institutional framework of the country's agriculture needs to be in place to apply the policies related to agriculture as well as to an appropriate market functioning that allows the agri-food chain to compete with EU companies on the internal market. Preparations observed so far have concentrated on adopting the legislative requirements of the EU *acquis* and first steps in the establishment of new institutions and the improvement of existing institutions. A large number of laws and regulations have been passed, but their effective implementation is only in a first stage in several areas. These will be discussed and evaluated below.

Serbia's law on competition policy has been renewed and was enforced in 2009. The law is compatible with EU's law. Abuse of dominant position on the market is prohibited, and in the reporting to the European Commission (Chapter 8 in Answers the EC) some cases are included to illustrate the interventions made by the Competition Authority to watch over malpractices. However, the administrative court, which is also responsible for enforcement of the law, was introduced into the system only in early 2010. The newly established institutions need time for building up capacities and expertise to do their work in a proper way.

A major concern with respect to institutions supporting agricultural market functioning is the unfinished process of restitution of properties, including agricultural land. After World War II private property was either nationalized or confiscated into state ownership.

Since the 1970s private landownership has been allowed but only to a maximum of 10 ha. The heritage of the past is a complex system of ownership and leasing, where property has not been returned to the original owners or their heirs. Currently, there is no law on restitution to settle this issue. A complicating factor is the incomplete registration of land in a cadastre. The lack of appropriate institutional arrangements on land property hampers the land market, and hence an efficient use of land resources. This effectively restrains agricultural development.

Serbia's farming sector will have to become much more efficient after membership, as it will encounter stronger competitive pressure from farmers in the rest of the EU. Increasing the capacity of the Serbian farming sector would be greatly facilitated by progress in an agricultural knowledge and information system, by appropriate priority setting for agricultural research, by adequate farm extension services and by improved linkages between farm and off-farm sectors. To date, the agricultural education, extension and research system suffers from a lack of financial means and coordination of activities based on long(er)-term programmes. What is particularly critical is that investments in research and education should be linked needs of knowledge – both for today and in the future -- at the level of both farmers and agribusiness. Knowledge creation, adoption and diffusion require an interplay of all actors involved. This in fact calls for further investment in human resources at all levels (universities, ministries, farmers, etc.). Strengthening the knowledge and innovation system also requires the preparation of the essential legal framework for protecting intellectual property rights. At present, progress made in harmonizing regulations of Serbia with those of the EU builds confidence, yet this is again in an early stage. As said in Chapter 3, the establishment of appropriate institutions for implementing the legislation in this area needs further attention.

The implementation of agricultural policy through the application of support and intervention measures is, of course, of major importance in the framework of EU accession. In the pre-accession period EU support in this area has been provided through the CARDS (2000-2006) and the IPA programme (part of EU's 2007-2013 budget programme). Currently, IPA is being implemented by the EU Delegation in Belgrade. The EU aims to transfer the management of IPA funds to national authorities under the Decentralized Implementation System (DIS) once the Serbian government develops the necessary capacity to assume this responsibility. Since the adoption of the DIS Road map, considerable progress has been made with the establishment of the required authorities. However, several issues were identified that might cause problems in meeting the requirements for DIS accreditation (Knezevic, 2010). One is an adequate number of sufficiently trained staff at country, regional and local level that is able to manage EU funds. A lack of expertise to complete technical documentation of large investment programmes has been a problem in using funds for infrastructural projects. Furthermore, Serbia lacks the culture and expertise to evaluate large investment projects. By applying for EU membership Serbia shows clear political support for the creation of DIS structures, which is further illustrated by the progress in establishing agricultural institutions related to preparations for the use of IPARD funds (see also Chapter 3).

The entire corpus of EU legislation (*acquis communautaire*) in agriculture, rural development, veterinary, phytosanitary issues and food safety policy comprises over 27,000 legislative acts with different degrees of compulsory compliance to the regulatory framework (regulations, directives, decisions, recommendations and opinions). Of that number, MAFWM has jurisdiction for over 9,000 regulations. It is therefore necessary to effectively coordinate between the relevant institutions in order to speed up the process of harmonization. Due to the fact that some institutions are not obligatory until the accession, it is important to purposefully map capacity and institution building and prioritize those areas that should be worked on in order to comply with EU requirements. Also, internal systems and the organization of MAFWM need to adapt to better respond to the requirements linked to the application of EU IPARD and ultimately CAP measures (Zivadinovic and Milovanovic, 2010).

11.6 Financial and economic consequences[1]

11.6.1 Introduction

In this section we present an estimation of the budgetary effects of Serbia entering the EU. In subsection 11.6.2 we present our assumptions for the estimate of the CAP expenditure, followed by our estimations of direct payments in Pillar 1 and support for rural development in Pillar 2. In subsection 11.6.3 we make some comments on the assistance from the EU structural funds in Serbia after accession. Subsection 11.6.4 concludes.

11.6.2 Assumptions for and estimations of the CAP expenditure

Assumptions for the estimate of the CAP expenditure

CAP expenditure can be divided into three groups:

Direct payments

- The CAP reform of 2003 introduced a single farm payment (SFP) in the EU-15 Member States and a single area payment scheme (SAPS) in most of the 12 new Member States. These payments are for the major part decoupled from production. When some coupling still exists, this will become decoupled as well in the coming years. The basis for the single payment scheme is the establishment of a maximum amount -- the national ceiling, which equals the total of direct aids paid in a historical period (generally 2000-2002) -- that each Member State can spend on direct aid (Jongeneel and Brand, 2011). Active farmers are granted entitlements to direct payments based on the amount of payments they received in the reference years.[2] The national envelop for new Member States has been agreed during the accession negotiations, and it is usually granted as a uniform hectare payment to all farmers.

Market support

- This expenditure covers export refunds, payments for intervention storage and other market interventions. Payments for market support fluctuate from year to year, depending on the market situation.

[1] This section builds largely on a background note by Ida Terluin and Ton de Kleijn, LEI, part of Wageningen UR.

[2] According to the Council Regulation (EC) 73/2009 farms must have a minimum size of 1 ha for being eligible for direct payments of Pillar 1. However, in order to take account of their specific agricultural structure, Member States are allowed to use another threshold. In practice, the threshold varies from 0.1 ha in Malta to 5 ha in Denmark, the UK and the Czech Republic (Annex VII of Council Regulation (EC) 73/2009). A minimum farm size is not included in our budget calculations.

Rural development support

- This support is given in the scope of the second pillar of the CAP. Payments for rural development are determined in advance by the EU Commission, the EU Council of Agricultural Ministers and the European Parliament, usually for a period of seven years.

Estimate of the decoupled direct payments

For the EU-15 Member States, direct payments granted in the reference period 2000-2002 are the base of the single farm payment. As these direct payments do not exist for Serbia, we calculate them based on the number of livestock and hectares and the related payments. As a reference period, we use the most recent years for which data are available, i.e. 2007-2009. As no production quota exists for Serbia, we assume that these are similar to the number of hectares and livestock in the reference period. In the estimate of the payments per hectare and livestock, our assumptions are as follows:

- Premiums per ha are calculated for the reference period 2007-2009. The premium for cereals and oilseeds amounts to 63 euro per tonne and that for protein crops to 72.50 euro per tonne. Based on the number of ha and the output per ha of each crop in the reference period, an average output per ha is calculated. For obtaining a premium per ha, this average output is multiplied by the premium per tonne (Table 11.1);
- For the calculation of the premiums for livestock, the premiums for the distinguished animals eligible for support are multiplied by the average number of animals in the reference period 2007-2009 (Table 11.2);
- The CAP reform of 2003 introduced a milk premium, which increases from 11.81 euro per tonne in 2004 to 35.50 euro per tonne in 2006. Afterwards, the milk premium is maintained at the level of 2006. For the calculation of a milk premium for Serbia, we multiplied the average amount of milk in the reference period 2007-2009 by the milk premium. The slaughter premium for dairy cows has already been included in the calculations of the premiums for livestock (Table 11.2);
- The premium for sheep is calculated by multiplying the average number of ewes in the reference period 2007-2009 by 25 euro per ewe (Table 11.2);
- The common market organization for sugar is self-financing. This implies that the calculation of the budget costs of accession of Serbia into the EU is exclusive of costs for a sugar premium.
- For wine and all other products not mentioned above (the so-called small products), no direct payments were used in the EU-15 in the years 2000-2002. These products were supported by market intervention. However, during the past years these payments for market intervention have been included in the single farm payment. In order to calculate a premium for wine, we divided the production of wine in the EU-15 in the reference period 2000-2002 by the expenditure for market intervention for wine. This results in a premium of 75 euro per hectolitre wine. A direct payment for wine in Serbia is calculated as the wine production multiplied by the wine premium.
- The payment for the small products is derived from the ratio of the expenditure for market intervention for wine in the EU-15 in the reference period 2000-2002 and those for the small products. In the EU-15, about 75 per cent of this expenditure was spent on the large products, 15 per cent on wine and 10 per cent on the small products. We assume that the ratio in this expenditure for Serbia would reflect the former EU-15

situation. Hence for calculating the direct payment for the small products, we multiply the direct payment for wine by about 66 per cent (10/15) (Table 11.3).

Table 11.1. Calculation of the direct payments for cereals, oilseeds and protein crops in Serbia.

	Cereals	**Oilseeds (a)**	**Protein crops (b)**
Average production, 2007-2009 (million tonne)	7,908	753	710
Average number of ha, 2007-2009 (1,000)	1,910	329	312
Average production per ha, 2007-2009 (tonne)	4.14	2.28	2.27
EU-15 premium per tonne, 2000-2002 (euro)	63	63	72.50
Direct payment per ha (euro)	260.82	143.64	164.58

Notes: (a) sunflower, soybeans, rapeseed and poppy seed; (b) beans, cowpeas, and peas.
Source: production and ha from Statistical Office of the Republic of Serbia and FAOSTAT: adaptation LEI.

Table 11.2. EU-premiums and Serbian reference production levels.

	EU-15 premium (euro)	**Reference production in Serbia, 2007-2009 (1,000 animals/1,000 tonne)**
Bulls	210	42
Heifers	150	16
Suckler cows	200	34
Slaughter premium cows	80	437
Slaughter premium calves	50	23
Ewes	25	1,328
Milk	35.50	1,566

Source: production from Statistical Office of the Republic of Serbia; adaptation LEI.

Table 11.3. Calculation of the direct payments for wine and small products in Serbia.

	Derived premium (euro)	**Production, 2007-2009 (1,000 hl)**	**Direct payments (million euro)**
Wine	75	168	12.6
Small products			66% of direct payments for wine: 8.4

Source: wine production from FAOSTAT; adaptation LEI.

The sum of the direct payments for the products above is given in Table 11.4. According to our calculations, the national envelope of direct payments for Serbia would amount to about 760 million euro per annum. This budget has about the same magnitude as that in Bulgaria (Table 11.5). This would imply a direct payment of 204 euro per ha of utilized agricultural area. This is close to the EU-12 average payment per ha of 190 euro. If this direct payment per ha in Serbia is phased in, as in the other new Member States, the payment per ha gradually increases from 51 euro in the first year after accession to the EU to 204 euro ten years later (Table 11.6).

Table 11.4. Estimated total direct payments in Serbia.

	Production, 2007-2009 (1,000 tonnes/animals/hl)	Premium per ha, tonne, animal or hl (euro)	Direct payments (million euro)
Cereals	7,908	260.82	498.2
Oilseeds	753	143.64	47.3
Protein crops	710	164.58	51.3
Bulls	42	210	8.8
Heifers	16	150	2.4
Suckler cows	34	200	6.8
Slaughter premium cows	437	80	35.0
Slaughter premium calves	23	50	1.2
Ewes	1,328	25	33.2
Milk	1,566	35.50	55.6
Wine	168	75	12.6
Small products			8.4
Total			760.7

Source: see Tables 11.1-11.3; adaptation LEI.

Table 11.5. Direct payments of Pillar 1 of the CAP in the EU-12 Member States and estimate for Serbia.

	National envelope with direct payments for Pillar 1 (million euro) (a)	Utilized agricultural area, 2007 (million ha)	Direct payment per ha (euro)
Latvia	146	1.77	83
Estonia	101	0.91	112
Romania	1,780	13.75	129
Lithuania	380	2.65	144
Poland	3,045	15.48	197
Slovak Republic	388	1.94	200
Czech Republic	909	3.52	258
Bulgaria	814	3.05	267
Slovenia	144	0.49	295
Hungary	1,319	4.23	312
Cyprus	53	0.15	366
Malta	5	0.01	498
EU-12	9,086	47.94	190
EU-27	45,813	172.49	266
Serbia	760.6	3.74	204

Note: (a) All data on the national envelop of MS are 2014 data except 2016 data for Bulgaria and Romania (when these countries are fully phased in). This is when we assume national envelops are not affected by the Multi-annual Financial Framework 2014-2020, although in reality this will be the consequence of applying the new MFF (see section 11.2).
Source: Direct payments Pillar 1 in the EU-27 from EU Council Regulation 73/2009 Annex VIII; agricultural area EU-27 from Eurostat FSS 2007; agricultural area Serbia from FAO; adaptation LEI.

Table 11.6. Estimate of phasing in of direct payments in Serbia.

Year	t	t+1	t+2	t+3	t+4	t+5	t+6	t+7	t+8	t+9
Phasing in percentage	25	30	35	40	50	60	70	80	90	100
Total direct payments (million euro)	190	228	266	304	380	456	532	608	685	761
Direct payment per ha (euro)	51	61	71	81	102	122	143	163	183	204

Estimate of the expenditure for market support

Payments for market support fluctuate from year to year, depending on the market situation. Payments for intervention storage are not part of the national envelop and will not be taken into account in this note. During the WTO negotiations in the Doha Round, the EU Commission has offered to eliminate all export subsidies on agricultural products by 2013. Although this Round is still not finished, we assume that this promise will be realised. As the accession of Serbia is after this date, we either do not take account of export refunds in this note.

Estimate of expenditure for rural development

By using the Instrument for Pre-accession Assistance (IPA), the EU provides financial assistance for countries on their way to EU membership (EC, 2011a). Currently, there are five candidate countries (Croatia, Iceland, the former Yugoslav Republic of Macedonia, Montenegro and Turkey) and four potential candidate countries (Albania, Bosnia and Herzegovina, Kosovo and Serbia). IPA consists of five components:

1. Transition Assistance and Institution Building
2. Cross-Border Cooperation
3. Regional Development
4. Human Resources Development
5. Rural Development

As a potential candidate country, Serbia receives financial assistance for components 1 and 2. For the years 2007-2013, about 1.4 billion euro are available for Serbia (Table 11.7). This is about 12 per cent of the total IPA budget 2007-2013 of 11.5 billion euro. However, Serbia receives no assistance for rural development, so this provides no guideline for our estimate of the EU expenditure for rural development after accession of Serbia to the EU. Therefore, we try to derive the expenditure for rural development in Serbia from the ratio of the direct payments in Pillar 1 and the EU support for rural development in the new Member States in the financing period 2007-2013 (Table 11.8). It appears that the share of the direct payments in Pillar 1 in the CAP budget 2007-2013 in the EU-12 amounts to 52 per cent and that of the EAFRD budget for rural development support in Pillar 2 to 48 per cent. However, the share of rural development support in the total CAP budget in Bulgaria and Romania is higher than the EU-12 average. This can partly be explained by the phasing in of the direct payments in these two countries, which lags behind those in the other new Member States, and which results in a moderate direct payment budget. In order to estimate the expenditure for rural development support for Serbia, we use the average of the shares of Pillar 2 in the CAP budget of the EU-12, Bulgaria and Romania: (48 + 51 + 59)/3 = 53 (Table 11.9). This results in a support for rural development in Pillar 2 in Serbia of 386 million euro p.a.[1]

Table 11.7. IPA support for Serbia, 2007-2013 (million euro).

	2007	2008	2009	2010	2011	2012	2013	Total
Transition Assistance and Institution Building	181.4	179.4	182.5	186.2	189.9	193.8	203.1	1,316.3
Cross-Border Cooperation	8.2	11.4	12.2	12.4	12.7	12.9	11.6	81.4
Total	189.6	190.8	194.7	198.6	202.6	206.7	214.7	1,397.7

Source: EC Commission at website http://ec.europa.eu/enlargement/potential-candidates/serbia/financial-assistance/index_en.htm.

Table 11.8. CAP budget for direct payments in Pillar 1 (P1) and rural development in Pillar 2 (P2), 2007-2013 (million euro).

	Total P1 + P2 (mln euro)	Direct payments P1 (mln euro)	Rural development P2 (EAFRD) (mln euro)	P1 as % of total	P2 as % of total
EU-15	297,447	246,242	51,205	83	17
Cyprus	380	217	163	57	43
Czech Republic	7,316	4,500	2,816	62	38
Estonia	1,209	494	715	41	59
Hungary	10,299	6,493	3,806	63	37
Lithuania	3,611	1,868	1,743	52	48
Latvia	1,766	725	1,041	41	59
Malta	97	20	77	21	79
Poland	28,269	15,039	13,230	53	47
Slovenia	1,612	712	900	44	56
Slovak Republic	3,892	1,923	1,969	49	51
Bulgaria	5,098	2,489	2,609	49	51
Romania	13,525	5,502	8,023	41	59
EU-12	77,074	39,982	37,092	52	48
EU-27	374,521	286,224	88,297	76	24

Source: budget direct payments 2007-2013 from Council Regulation 1782/2003 (consolidated version - 5 August 2006) and from 'Threat of SFP cuts rises as NMS accede', Agra Europe Weekly, 12 January 2007; P2 (= EAFRD) budget 2007-2013 from EC (2007), EU support for rural development 2007-2013; Pre-allocated funding under Heading 2 'Natural Resources' of the Financial Framework, Brussels; adaptation LEI.

[1] Considering the low GDP/capita in Serbia, the country will be earmarked as a 'convergence' region. That means for rural development (Pillar 2) an EU co-financing of 75% for Axes 1 and 3 and 80% of axes 2 and 4 (Council Regulation (EC) 1698/2005).

Table 11.9. Estimate of the budget for rural development support for Serbia.

	Budget direct payments Pillar 1, 2007-2013 (million euro)	Budget rural development support Pillar 2, 2007-2013 (million euro)	Share P1 in CAP budget (%)	Share P2 in CAP budget (%)	Serbia direct payments with phasing in during first 7 years (million euro)	Estimated rural development support Pillar 2 in Serbia for 7 years by using share P2 in CAP budget (million euro)	Estimated rural development support Pillar 2 in Serbia p.a. (million euro)
EU-12	39,982	37,092	52	48	2,358	2,187	312
Bulgaria	2,489	2,609	49	51	2,358	2,472	353
Romania	5,502	8,023	41	59	2,358	3,438	491
Average			47	53	2,358	2,699	386

Estimate of the CAP budget costs of accession of Serbia to the EU

In section 11.2 we estimated a national envelop for direct payments in Pillar 1 and budget expenditure for support for rural development in Pillar 2 for Serbia. In this section we integrate the budgets for the separate parts and put them in an EU perspective.

The estimated budget for direct payments in Pillar 1 for Serbia would amount to 761 million euro p.a. and that for support for rural development in Pillar 2 to 386 million euro p.a. (Table 11.10). If we take account of the usual phasing in of the direct payments of Pillar 1, then the total CAP budget for Serbia for the first seven years after accession is about 5 billion euro. This equals about 1.4 per cent of the CAP budget for the EU-27 in the period 2007-2013. This estimated CAP budget for Serbia for the first 7 years after accession is rather close to the budget for Bulgaria in the period 2007-2013 (Table 11.8).

Table 11.10. Estimate of CAP budget costs of accession of Serbia to the EU.

	Payments to Serbia p.a. (million euro)	**Payments to Serbia in a period of 7 years with phasing in of direct payments in P1** (million euro)	**EU-27 CAP budget, 2007-2013** (million euro)	**Payments to Serbia as share of EU-27 CAP budget** (%)
Direct payments Pillar 1	761	2,358	286,223	0.8
Support for rural development in Pillar 2	386	2,699	88,297	3.1
Total	1,146	5,057	374,520	1.4

11.6.3 Assistance to Serbia from the EU structural funds

Apart from the common agricultural policy, another main policy field of the EU refers to cohesion policy for growth and jobs in order to strengthen cohesion among the regions in the EU. The funds for the structural policy are usually granted for programming periods of seven years. The next programming period will cover the years 2014-2020. The structural funds are the European Fund for Regional Development (EFRD), the European Social Fund (ESF) and the Cohesion Fund. Especially regions with a GDP/capita of less than 75 per cent of the EU-27 average are supported by the EU structural funds. GDP/capita in all NUTS2 regions in Serbia is below this threshold (see Chapter 2). In order to safeguard the absorption capacity of the granted funds to a country, the support may not exceed 4 per cent of the GDP of the country.

In terms of GDP/capita, Bulgaria is close to Serbia. In the current programming period 2007-2013, Bulgaria receives nearly 6.9 billion euro from the structural funds (i.e. heading 1b: Cohesion for growth and jobs) (EC, 2011b). This is nearly 2 per cent of the total EU budget for cohesion policy (348 billion euro). A rough estimate for the assistance from the structural funds for Serbia would be that Serbia gets the same amount of support as Bulgaria in the period 2007-2013 (Table 11.11). This is equal to about 979 million euro p.a.[1]

Table 11.11. Rough estimate of structural fund assistance for Serbia.

	2007-2013	**per year**
Assistance to Bulgaria from cohesion policy (million euro)	6,853	979
	Seven-year period	**per year**
Estimate of assistance to Serbia from cohesion policy (million euro)	6,853	979

Source: Assistance to Bulgaria from EC (2011b).

Serbia's contribution to the EU budget

Serbia's contribution to the EU budget is based on the assumption that this will be on the basis of the country's GDP. The general rule is that a member state's contribution amounts to 1.14 per cent of GDP at the most.[2] Starting from the 2009 level of 30.0 billion euro (see Chapter 2), a registered 2 per cent growth for 2010 and an assumed 3 per cent growth for 2011, plus 5 per cent annual growth for five to seven years onwards would bring Serbia's GDP in the range of 40.2-44.3 billion euro at the time of accession. This would then result in a contribution of 460-505 million euro (at 2016-2018 values).

11.6.4 Concluding remarks: total effects

Table 11.12 summarizes the budgetary transfers arising from Serbia's EU accession in five to seven years. Taking into account this time frame, and the options of phasing in or not phasing in, gives the country a net budget revenue of 1.1 to 1.67 billion euro if Serbia becomes a member within 5 years. If Serbia enters in 7 years, the net budget revenue for Serbia will be 45 million euro less, as the country's contribution to the EU budget will be

[1] The EU co-financing ceilings for public expenditure amount to 75% for the ERDF and the ESF and 85% for the Cohesion Fund.

[2] Actual contributions are based on Gross National Income (GNI). However, in the MFF proposals the Commission suggests switching to an own-resource system that is based on a financial transactions tax and a new VAT resource (EC Communication, 2010: 8).

higher. The amounts are about 2.6 to 3.7 per cent of Serbia's GDP over five or seven years, respectively, which is a significant injection into the economy.

Budgetary transfers will play an important role for economic growth as the means generated will be used for local consumption and investment, which adds to further enhancement of the country's economy. Next, the consequences of EU membership for Serbia are the longer-term economic benefits that arise from aspects such as more opportunities for trade, greater transparency in governance matters, lower corruption and a clear direction of policy change.

Table 11.12. Summary of budgetary transfers arising from Serbia's accession

Budget item	**Amount (million euro per year)**	
	MS in 5 years	**MS in 7 years**
CAP first pillar	190/760 (a)	190/760 (a)
CAP second pillar	386	386
Structural and cohesion policy	979	979
Contribution of Serbia	460	505
Net budget revenue	1,095/1,665	1,050/1,620

Note: (a) lower versus higher end of the range is with or without phasing in of the direct payment in the first year of EU membership.

11.7 Consequences for the EU of the enlargement to Serbia

While the impact of EU membership on the Serbian economy is expected to be substantial, consequences for the countries that are already EU members at the time Serbia joins are assumed to be relatively small. This is based on the relatively small size of the Serbian economy compared to EU-27 (or 27+), the rather limited transfer of money from EU members to Serbia, and the small size of the Serbian agricultural sector, including the upstream and downstream industries, in the EU context. Given these aspects the entrance of Serbia is not expected to lead to policy reforms as such.

Serbia's economy is (almost) the size of Bulgaria's, and in terms of GDP only 0.25 per cent of the EU-27 (see Chapter 2). Serbia's population is 1.4 per cent of EU-27's total and its per capita purchasing power is far below the Union's average. Even with a substantial economic growth that exceeds the EU average in the years to come, Serbian economic development will have so little weight that it will hardly affect the EU economy. Of course, the increase in purchasing power in Serbia may provide some additional market opportunities for EU industries. It is beyond the scope of this report to explore in more detail these consequences for sectors other than agriculture and food; but, again, given the relatively small size of Serbia's economy, the general economic effects on EU-27 because of these opportunities are expected to be rather insignificant.

Budgetary transfers to Serbia are estimated to be 1.1 to 1.6 billion euro (see section 11.6). This flow of money implies a (net) transfer between member states and hence has a redistributive effect from EU-27 to Serbia. How strong this effect will be depends on the impact this transfer of money will have on the rate of economic growth of both the transferor (in this case the EU-27) and the transferee (Serbia). Looking at the possible effects on the EU economy, the transfer of income (which is less than 0.1 per cent of the EU's GDP) is so small that any impact on the EU's economic growth is expected to be negligible. (even positive if money is spent on imports from EU).

Serbia is an important cereal producer in the Balkan region, as well as a significant producer of several kinds of fruits. However, compared to the EU-27 total output, these crops do not add significant volumes to the EU market, and will not affect the trade flows within the EU or among the EU and third countries to any significant extent. Therefore, no implications are expected for the EU single market organization in terms of reforms of intervention mechanisms (such as private storage, public storage, special measures, intervention safety net) that would be necessary because of Serbia's accession.

11.8 Concluding remarks and some recommendations to take up challenges of future EU integration

This chapter has evaluated the possible consequences for the Serbian agricultural sector if the country joins the EU in five to seven years. Of course, the evaluation is highly conditional on our assumptions on general economic growth, exchange rate developments and EU policy developments. Based on our assumptions underlying the budgetary consequences, Serbia's joining the EU would imply an annual transfer of 1.05 to 1.67 billion euro. This implies a significant financial injection into the Serbian economy. Economic consequences for the present EU members are minor.

EU membership will have major implications for Serbia's agricultural policy and the sector's structure. At present, Serbia's agricultural policy differs significantly from EU's CAP, as the instruments which are used and the criteria for and level of support are different (see Chapter 7). Subsequently, aligning with EU policy will have significant distributional effects, not only among farmers (as criteria for and levels of support will change), but also between the agricultural and non-agricultural parts of the economy. Furthermore, the full implementation of the *acquis* on agriculture, food safety, veterinary and phytosanitary issues -- and the environmental measures to avoid adverse effects from agricultural activities -- require investments at farm and firm level. IPARD and (when an EU member) rural development funds are available to adapt to the requirements set by the *acquis*. Yet, parts of the primary and processing sector will find themselves too small to make these investments economically. Consequently, the sector will face a restructuring and consolidation process. While this process increases efficiency and, hence, competitiveness of the sector -- and thus is beneficial from the sector's perspective -- there are possibly substantial social consequences...such as finding alternative employment for those leaving the agricultural sector, which may be problematic. Economic growth that creates employment opportunities in other sectors is an important condition for a smooth process of structural change in the agricultural sector.

Serbia faces many challenges should the country become an EU member in the medium term. A major condition for EU membership is the adoption and implementation of the set of rules, laws and regulations that is the *acquis communautaire* of the European Union. Serbia has been in a process of substantial reforms of institutions and legislation since the political changes in 2000. In that process the country's administration has been adapting legislation in line with the EU, and that process has accelerated in the most recent years. Nevertheless, the analysis presented in Chapter 3 indicates that fundamental reform of laws and institutions has been slow, especially with respect to land, cooperatives, advisory services, the food safety system and so on, due to a lack of inter-ministerial cooperation and an extensive overlapping of the responsibilities between different institutions. Substantial structural reforms in the sector were delayed as a result. Speeding up the process of adopting legislation and establishing the necessary institutions to implement the legislation are both required to provide the agri-food chain a well-founded legal and institutional context in which to do business.

In view of EU membership, the Serbian government has adopted many administrative arrangements, whereas administrative and social capacities to implement these arrangements are still deficient. Enhancing human resources of the administration is certainly a priority, as is better coordination between the various governmental bodies. Several Twinning projects have been initiated to support capacity improvements, such as in the area of food safety and farm data (linked to FADN). Intensifying the process of capacity building is recommended, as improving competences of the public administration is a crucial element in the process of preparing for EU membership in the years to come.

In the medium term the CAP will remain the principal framework for the agricultural policy in the EU. Therefore, aligning agricultural policies with the CAP will facilitate later accession to the EU. In bringing its policies in line with those of the EU, Serbia needs to be well informed about the way the CAP evolves. Such information is important in order to evaluate the possible impact of policy changes on the sector's development, and from there to design appropriate policy answers to prepare the sector for the effects identified.

The CAP moves to rural development (second pillar) measures. That by itself is a reason to give ample attention to that policy area in Serbia's process of preparing for EU accession. Moreover, effective rural development programmes are also needed to ensure that the agricultural sector's modernization is balanced and equitable. Serbia's agricultural sector has a dual structure with many small-scale, subsistence farms, and regional differences in economic development are considerable. Restructuring and modernization of the agri-food sector, therefore, may have significant regional economic effects. Thus it is important, to explore the non-agricultural potential of rural areas and design territorial, rural development strategies. Local authorities and stakeholders should be involved in designing these programmes.

A major task for the Serbian government (MAFWN) is to provide sufficient public services that help the agri-food sector to prepare for EU membership. As said above, the Serbian agri-food supply chain may expect increasing competition if Serbia joins the internal market. Advisory (extension) services, agricultural information services and agricultural education and research are public services with a proven valuable impact in helping adapting the agribusiness to changing policy and market conditions. Currently these services fall short of what the sector needs. Increasing public investments in these areas would enhance the sector's capacity to stand the increasing competitive pressures it will encounter when Serbia is a member of the EU.

Literature and websites

EC, *Serbia 2010 Progress Report*. Commission Staff Working Document (COM(2010)660). Brussels, 9 November 2010.

EC, *The Common Agricultural Policy towards 2020. Meeting the food, natural resources and territorial challenges of the future*. Communication from the Commission to the European parliament, the European Economic and Social Committee and the Committee of the regions, Brussels 18/11/2010, COM(2010)672 final. 2010.

EC, *Instrument for pre-accession assistance (IPA)*. European Commission, Brussels, 2011a. Available at: www.ec.europa.eu/enlargement/how-does-it-work/financial-assistance/instrument-pre-accession_en.htm

EC, *Pre-allocation by member state for certain EU-support 2007-2013*. 2011b. Available at: http://ec.europa.eu/budget/biblio/documents/fin_fwk0713/fin_fwk0713_en.cfm#aii

IMF, *Economic Outlook, June 2011*. Washington, IMF.

Jongeneel, R. and H. Brand, 'Direct income support and cross-compliance.' In: Oskam, A., G. Meester and H. Silvis, H. (eds.), *EU policy for agriculture, food and rural areas.* Wageningen Academic Publisher, Wageningen, 2011.

Knezevic, I., *Absorption capacity of Serbia for use of EU funds: Practical lessons from Slovakia.* The Pontis Foundation and the Center for Democracy Foundation, Belgrade, 2010.

Lampietti, J., D. Lugg, P. van der Celen and A. Branczik, *The Changing Face of Rural Space Agriculture and Rural Development in the Western Balkans.* World Bank, Washington DC, 2009.

Project preparation and technical assistance. *Facility to Reinforce Administrative Capacity in Serbia.* http://ipa2007ppf.org

Zivadinovic, B. and M. Milovanovic, *Guide through EU policies - Agriculture (in Serbian.* Belgrade, 2010.

Subject Index

Acronyms

AC-2	Bulgaria and Romania ('accession countries')
ALC	Association of Leasing Companies
AMS	Aggregate Measure of Support
BSE	Bovine Spongiform Encephalopathy
BTB	Bovine Tuberculosis
CAP	(EU) Common Agricultural Policy
CBD	Convention on Biological Diversity
CEFTA	Central European Free Trade Agreement
CET	Common External Tariff
CIS	Commonwealth of Independent States
CMO	Common Market Organization
CO_2	Carbon dioxide
CNCT	Combined Nomenclature of Customs Tariffs
CPI	Consumer Price Index
CSF	Classical Swine Fever
CUS	Cooperative Union of Serbia
DAP	(Serbian MAFWM) Directorate for Agrarian Payments
DIS	Decentralized Information System
DOS	Democratic Opposition of Serbia
DREPR	Danube River Enterprise Pollution Reduction
DS	Reformist Democratic Party
EBRD	European Bank for Reconstruction and Development
EC	European Commission
ECP/GR	European Cooperative Program for Crop Genetic Resources
ECA	Europe and Central Asia
EEA	European Environment Agency
EAFRD	European Agricultural Fund for Rural Development
EFRD	European Fund for Rural Development
EFTA	European Free Trade Agreement
EIONET	European Information and Observation Network
EPPO	European Plant Protection Organization
ESA	European system of national accounts
ESF	European Social Fund
EU	European Union
EU-12	Bulgaria, Czech Republic, Cyprus, Estonia, Hungary, Latvia, Lithuania, Malta, Poland, Romania, Slovakia, Slovenia
EU-15	Austria, Belgium, Denmark, Finland, France, Germany, Greece, Ireland, Italy, Luxembourg, Netherlands, Portugal, Spain, Sweden, United Kingdom
EUNIS	European Nature Information System
FADN	Farm Accountancy Data Network (EU)
FAO	Food and Agriculture Organization
FD	*Flavescence dorée*
FDI	Foreign direct investment
FFV	Fresh fruit and vegetable sector

FISIM	Financial Intermediation Services Indirectly Measured
FMD	Foot and Mouth Disease
FRY	Federal Republic of Yugoslavia
FTA	Free Trade Agreement
FVO	(EU) Food and Veterinary Office (DG SANCO)
GAO	Gross agricultural output
GATT	General Agreement on Tariffs and Trade
GDP	Gross domestic product
GHG	Greenhouse gas
GMO	Genetically modified organism
GNI	Gross National Income
GoRS	Government of the Republic of Serbia
GVA	Gross Value Added
HACCP	Hazard Analysis and Critical Control Points
HNV	High Natural Value
IACS	(EU) Integrated Administrative Control System
ICT	Information and Communication Technology
ICTY	International Criminal Tribunal for the former Yugoslavia
IIRP	(Serbian MSTD) Integral and Interdisciplinary Research Programme
ILO	International Labour Organization
IMF	International Monetary Fund
IPA	Instrument for Pre-accession Assistance
IPARD	Instrument for Pre-accession Assistance for Rural Development
ITA	Interim Trade Agreement
IPM	Integrated Pest Management
IPPC	International Plant Protection Convention
IT	Information Technology
IUCN	International Union for Conservation of Nature and Natural Resources
LAG	Local Area Group
LEI	Agricultural Economics Research Institute (Landbouw-Economisch Instituut)
LFA(s)	Less Favoured Area(s)
LFS	Labour Force Survey
LPIS	(Serbian MAFWM) Land Parcel Identification System
LSMS	Living Standards Measurement Survey
LTAA	Long term annual average
MAFWM	(Serbian) Ministry of Agriculture, Forestry and Water Management
MESP	(Serbian) Ministry of Environment and Spatial Planning
MSTD	(Serbian) Ministry of Science and Technological Development
MFF	(EU) Multi-annual financial framework
NACE	(EU) Nomenclature des activités économiques dans la Communauté Européenne
NATO	North Atlantic Treaty Organization
NBS	National Bank of Serbia
n.d.	no date
NES	(Serbian) National Employment Service
n.e.s.	not excluded specifically
NMS	(EU) New Member States as of 1 May 2004
NPI	(Serbian) National Programme for Integration
NPRD	National Programme for Rural Development
NSDA	(Serbian) National Sustainable Development Strategy
NUTS	(EU) Nomenclature d'unités territoriales statistiques

OECD	Organization for Economic Cooperation and Development
OIE	Office International des Epizooties
PAP(s)	Processed agricultural product(s)
PIP	Project Implementation Plan
PPP(s)	Purchasing Power Parity(ies)
PWMC	Public Water Management Autority
RDS	Republic of Serbia dinar
RGZ	Republican Geodetic Authority
SAA	Stabilization and Association Agreement
SAP	Socialist Autonomous Province
SAPS	(EU) Single Area Payment Scheme
SAPARD	(EU) Special Accession Programme for Agriculture and Rural Development
SEPA	Serbian Environmental Protection Agency
SFP	(EU) Single Farm Payment
SITC	Standard International Trade Classification
SME(s)	Small and Medium Enterprise(s)
SORS	Statistical Office of the Republic of Serbia
SPS	Sanitary and Phytosanitary
SR	Socialist Republic
TB	Tuberculosis
TRQ(s)	Tariff Rate Quota(s)
UAA	Utilized Agricultural Area
UPOV	Union of Protection of New Varieties of Plants
UN	United Nations
UNDP	United Nations Development Programme
UNESCO	United Nations
UNFCCC	United Nations Framework Convention on Climate Change
USD	United States (of America) dollar
USDA	United States (of America) Department of Agriculture
USDA/FAS	USDA Food and Agriculture Survey
VAT	Value Added Tax
WAHIS	World Animal Health Information System
WHO	World Health Organization
WIPO	World Intellectual Property Organization
WTO	World Trade Organization

Selected Bibliography

Legislation

Codice del lavoro, G. Fava, P.A. Varesi, IPSOA, 2022.
Codice del lavoro, P. Pucci, F. Toffoletto, IlSole24Ore, Pirola, 2022.
Commentario breve alle leggi sul lavoro, R. De Luca Tamajo, O. Mazzotta, Cedam, Padova, 2022.
Codice del Lavoro, M.T. Carinci, Giuffrè, Milano, 2022.

Law Reports (General)

Foro Italiano (Foro It.) and Repertorio Foro Italiano, Bologna – Rome, 1876ff. Giurisprudenza Italiana (Giur. It.) and Repertorio Giurisprudenza Italiana, Torino, 1862ff.

Special Periodicals (Also Reporting Cases)

Giornale di Diritto del Lavoro e delle Relazioni Industriali, Angeli, Milan, 1978ff.; Diritto delle Relazioni Industriali (Dir. Rel. Ind.), Giuffrè, Milan, 1991ff.; Il Diritto del Lavoro (Dir. Lav.), Rome, 1927ff.; Lavoro e Diritto (Lav. Dir.), il Mulino, Bologna, 1986ff.; Massimario di Giurisprudenza del Lavoro (Mass. Giur. Lav.), Rome, 1928ff.; Orientamenti di Giurisprudenza del Lavoro (Orient. Giur. Lav.), Milan, 1951ff. Rivista di diritto del lavoro (Riv. Dir. Lav.), Giuffrè, Milan, 1948ff.; since 1982. Rivista Italiana di Diritto del Lavoro (Riv. It. Dir. Lav.);. Rivista Giuridica del Lavoro e della Previdenza Sociale (Riv. Giur. Lav.), Rome, 1948ff.

General Periodicals, but of Great Importance to Labour Law

Le nuove leggi civili commentate (N.L.C.C.), Cedam, Padua, 1977ff. Rivista trimestrale di diritto e procedura civile (Riv. Trim. Dir. Proc. Civ.), Giuffrè, Milan, 1947ff.

General Books

Ballestrero M.V., De Simone G., *Diritto del lavoro*, Giappichelli, Torino, 2022.
Candian A., Squeglia M., Tursi A., *Manuale della previdenza complementare*, Giappichelli, Torino, 2022.
Carinci F., De Luca Tamajo R., Tosi P., Treu T., *Diritto Sindacale*, UTET, Turin, 2018 (8th ed.).

Selected Bibliography

Carinci F., De Luca Tamajo R., Tosi P., Treu T., *Il rapporto di lavoro*, UTET, Turin, 2022 (11th ed.).
Cella G.P., Treu T., *Relazioni industriali e contrattazione collettiva*, Mulino, Bologna, 2009.
Cinelli M., *Diritto della previdenza sociale*, Giappichellli, Torino, 2022.
Del Punta R., *Diritto del lavoro*, Giuffrè, 2022 (14th ed.).
Magnani M., *Diritto del lavoro*, Giappichelli, Torino, 2021.
Magnani M., *Diritto sindacale*, Giappichelli, Torino, 2021.
Mazzotta O., *Diritto del Lavoro*, Giuffrè, Milano, 2022.
Persiani M., D'Onghia M., *Diritto della sicurezza sociale*, Giappichelli, Torino, 2022.
Vallebona A., *Istituzioni di diritto del lavoro*, II, *Il rapporto di lavoro*, Cedam, Padua, 2021.

Index

The numbers here refer to paragraph numbers.

Index

www.ingramcontent.com/pod-product-compliance
Lightning Source LLC
Chambersburg PA
CBHW071432230926
51846CB00001B/203
9789403504872